RESCUE OF THE 1856 HANDCART COMPANIES

Charles
Redd
Monographs
in Western
History No.11

RESCUE OF THE 1856 HANDCART COMPANIES

REBECCA BARTHOLOMEW • LEONARD J. ARRINGTON

Brigham Young University
Charles Redd Center for Western Studies

The Charles Redd Monographs in Western History are made possible by a grant from Charles Redd. This grant served as the basis for the establishment of the Charles Redd Center for Western Studies at Brigham Young University.

Center editor: Howard A. Christy

Cover: Gary E. Smith
Handcart Pioneers on the Trail in Wyoming, *1978*
Oil on canvas, 30" × 54"
Collection of Leonard J. Arrington

The painting on the cover conveys a realistic impression of the tragic historical scene of 1856. Commissioned privately by authors Leonard Arrington and Davis Bitton in 1978, it was used as a frontispiece illustration in the first edition of this monograph.

Award-winning Mormon artist Gary E. Smith has painted many works now exhibited at BYU, the LDS Church Museum, the Utah State Alice Art Collection, and elsewhere; he has also created murals in Salt Lake City, Murray, and Provo. His imaginative semiabstractionist works deal particularly with Mormon themes reinterpreted away from traditional prototypes.

Library of Congress Cataloging-in-Publication Data

Bartholomew, Rebecca, 1945–
Rescue of the 1856 handcart companies / Rebecca Bartholomew and Leonard J. Arrington.
p. cm.—(Charles Redd monographs in western history; no. 11)
Originally published: Provo, Utah: Brigham Young University Press, c1981.
Includes bibliographical references and index.
ISBN 0-941214-04-4 (softbound)
1. Overland journeys to the Pacific. 2. Mormon Church—West (U.S.)—History. 3. Handcarts—West (U.S.)—History—19th century. 4. West (U.S.)—History—1848–1950. 5. Frontier and pioneer life—West (U.S.). I. Arrington, Leonard J. II. Title. III. Series.
F593.B26 1993
917.804'2—dc20 92-42327
CIP

Charles Redd Center for Western Studies, Provo, Utah 84602

Revised edition, 1993
Printed in the United States of America

Distributed by Signature Books, 564 W. 400 N., Salt Lake City, Utah 84116-3411

Contents

William Henry Jackson, *Handcart Pioneers,* detail of print. Photograph courtesy of Utah State Historical Society.

Introduction

In 1855 Brigham Young wrote from Great Salt Lake City to Edmund Ellsworth, a son-in-law in England:

> We are very ancios to have a company got up in England to cros the planes with hancarts. I due beleve that I could bring a company across, without a team and beet eny ox trane if I could be there myself. Would you like to try it?[1]

This was not Young's first mention of handcarts, but it was the first time the proposal would be acted upon. The handcart scheme was a "last resort," employed only when drought and other economic adversities had withered resources until the church could no longer afford to outfit the poor with Chicago wagons and ox teams.[2] Yet mission leaders were writing home from Europe that twenty thousand converts "clamored" to come to Zion; officials could go nowhere without being teased by the poor to "let them come."[3] So Brigham Young's letter to Edmund Ellsworth continued:

> If we can have our emegration come to the eastern citys and the northan rout, it will be much relieve [to] our Brethrern from sickness and deth which I am very ancious to due. There is a raleway from New Yourk City to Iowa City and will cost onley about 8 dollars for the pasedge. Then take han[d]carts and there little luggedge with a fue good milk cowes and com on till they are met with teams from this place, with provisions &c—

The handcart proposal was received by the European Saints with jubilation. Most LDS officials, too, responded enthusiastically and outdid themselves in accommodating those who wished to emigrate immediately. Not one but five companies, comprising nineteen hundred people, sailed from Liverpool through the winter and spring of 1855–56, all converging on Iowa City.

From Iowa City the 1856 immigration turned into the mixture of success and tragedy which has been fully documented by LeRoy and Ann Hafen, with the tragedy eloquently retold by Wallace Stegner.[4] When all but the first company arrived in Iowa, little was ready for them. Though (as Franklin D. Richards, president of the European Mission, later swore) "hard thinking, hard working, and doing the best we could" went into the planning, deliberations had not ensured sufficient stockpiles of hardwood, iron, and canvas for carts and tents.[5] From the Utah end, only one of the proposed supply stations had been established. In spite of this, three companies left Iowa in early June and arrived in Utah in less time and with casualties equal to or fewer than the typical wagon train. True, they were short of provisions all the way and would have suffered severely if supply trains had not reached them three hundred miles east of Salt Lake City. But immigrants and church leaders considered the experiment a success.

Error and misfortune dealt the two remaining handcart companies, the so-called Willie and Martin companies, "the worst disaster in the history of Western migration."[6] Church historian Andrew Jenson concluded, "after the lapse of thirty-seven years, and weighing the matter calmly in well balanced minds," that no one was "specially to blame"; perhaps certain elders in England had responded "too kindly" to some Saints overanxious to reach the mountains.[7] There had been difficulty procuring boats so that the largest company had not sailed until the end of May. Long waits in Iowa City and Florence, Nebraska, had been necessary as agents scrambled to outfit the unexpectedly large and late horde. From the start the Martin company had inherited a heavier share of infants, widows with large families, sick, and aged.[8]

When warned that it was too late to cross the Rockies that year, company leaders, encouraged by immigration officials, made a decision which in hindsight seems almost cavalier: They would start anyway, trusting a divine hand to temper the weather and other realities. Most of the immigrants were eager to go, although a few followed because there seemed little choice—otherwise they must winter in a territory hostile to Mormons, where jobs were scarce and survival depended upon homesteading skills they had not learned in the factories. Like generals planning a battle, knowing their attack might be stalled by an unreasonable storm but counting on normal weather, immigration leaders made the decision to go on.[9]

Even with good timing and luck, travel by handcart was a risk. Apostle John Taylor was skeptical from the beginning, and Mormon historian B. H. Roberts later described the whole idea as "possible but not feasible." The Willie and Martin companies soon found it impossible. Once on the trail, carts broke down, cattle stampeded, provisions ran out, and the most violent winter to hit the region in many years descended before they were out of the Black Hills—a month before usual snowfall. For once the tragedy could not be blamed on mobs. Given the weather, the causes lay in their own incaution and mismanagement, the responsibility for which lay equally with immigrants and mission leaders and, to a lesser degree, authorities in Utah. The tragedy is heightened by the fact that a decision that would have been applauded if the weather patterns had been normal came to be regarded as unthinking and callous, if well intended.

A lesser-known aspect of the handcart immigration, however, provides one of the most satisfying episodes in Mormon history. The rescue effort mounted by Brigham Young before anyone in Utah suspected the critical situation of the companies, the munificent response of communities throughout the territory, and the courage and endurance of the rescue parties make a heartening story. In many immigrant journals, memory of the rescue and the welcome in Salt Lake City dominates other experience; the drama and timeliness of the deliverance is dwelt on far more often than grief and disillusionment over the predicament. Those embittered were in the minority; and even these, decades later, wrote emotionally, gratefully, of the heroic men who saved them from starvation and death. The rescue is the subject of this essay.[10]

On 4 October 1856, the Willie Company, which had been short of teams when Franklin D. Richards camped with them in mid-August, was four days west of Fort Laramie, halfway to the "Upper Crossing" where they would leave the Platte River and strike overland to the Sweetwater. Each morning heavy frost made the walking unpleasant for those barefoot or in tattered shoes. Roads were deeply rutted from storms, putting even more strain on the frail carts. Members had just been assigned half rations and would soon have this amount halved again. Yet only a few in the camp realized how truly serious their condition was. The others, as sixteen-year-old Emma James later wrote, "could look toward the west and see the snow-capped mountains in the distance" and remain cheerful. They were still sturdy, though getting a little hungry.[11]

Captain Willie, Levi Savage, others of the subcaptains, and a few fathers had made calculations and knew that supplies were insufficient to take them to South Pass, the first place they could expect help from the Salt Lake Valley. Savage noted matter-of-factly in his journal the weak state of the teams and how the old people were failing fast. Even if the weather continued mild in daytime, some of their people were in trouble; any more thunderstorms or, worse, any snow would have fearful results.[12]

The Martin company was eight days behind them, still forty miles east of the fort. At Laramie they too were put on rations of two cups of flour per adult per day, so some visited the fort to trade jewelry, utensils, and heirlooms for cornmeal, beans, and bacon. Up to now most had found the journey pleasant, even though they had disposed of bedding to lighten the cartloads and make room for foodstuffs. Beyond Fort Laramie things would get rough and then grim, but for now only those with experience crossing the Rockies knew enough to worry.[13]

Organization of the Rescue

On 4 October, Elder Franklin D. Richards and his party of missionaries returned to Great Salt Lake City after absences of from two to four years. They had traveled by horse and carriage from Florence in about six weeks, checking on the welfare of the handcart companies as they overtook them. Until Richards's report, it was assumed in Salt Lake that the third handcart company which had arrived two days earlier was the last of the 1856 immigration. Brigham Young had already sent routine supply wagons out to several companies he knew to be on the road: Abraham Smoot's church supply train, the Walker and Gilbert & Gerrish private merchant trains, an immigrant company from Texas, and the church herd with accompanying wagons.[14] Now he was surprised to learn that two full handcart companies (whose numbers exceeded the total of the first three companies combined) plus a wagon company carrying luggage for the year's immigration were still on the trail—over twelve hundred souls.[15]

Richards was not a worrier. He had "this faith and confidence towards God...that [the immigrants'] path may be freed from suffering more than they can bear."[16] Rather as an afterthought he had purchased thirty-seven buffalo robes for Willie's company to claim at Fort Bridger, and he expected the Martin company to get in by their own steam the last of November. Of course the companies would need flour and teams, especially Brother Willie; but Richards's real concern seemed to be for a wealthy benefactor of the English Saints who was traveling with John A. Hunt's wagon company. "You will like Brother Tennant," he told President Young. Young responded, "I will like him if he is a saint."[17]

Perhaps Richards's attitude was not casual but simply confident that supplies could reach the companies before bad weather did. At any rate, Brigham Young's concern was immediate and more pronounced. He did not delegate the fact finding to his Presiding Bishopric but himself called a meeting, the very evening of Richards's arrival, to ascertain the location and condition of the handcart companies. At this meeting were the First

Presidency; eleven returning missionaries; managers of the tithing store, church store, and church herds; several local bishops; commander of the territorial militia Daniel H. Wells; and several clerks. "The object of my wanting the brethren here is to find out what we need to do tomorrow," Young told the group. With a semiannual church conference about to convene, a call for teams and supplies could be made immediately. "It is a day for calling on the Bishops."[18]

The minutes of this meeting (although no doubt made from sketchy notes) reveal personalities and the manner of making decisions. To Richards and the missionaries Young said, "Suppose we take the first company we come to as we go back, and we will see what you think they will need." One company nearest the valley was A. O. Smoot's freight train. Smoot had sent a letter via the missionaries calling for "40 yoke of cattle, 20 good teamsters, 3 or 4 beeves, 6 good wagons." Slightly ahead of him was the small Porter Rockwell train and tagging behind him for protection was the Gilbert & Gerrish store train (Walker had presumably wintered at Fort Laramie). Far behind was Willie's fourth handcart company. The minutes continue:

> PREST YOUNG: Next Company
>
> JAMES FERGUSON: Brother Willy is in want of teams—has one yoke of oxen to a wagon
>
> PREST YOUNG: He ought to have 10 yoke of cattle
>
> F D RICHARDS: All the help that can be sent will help them to come on so much faster. It wants good active teams
>
> JAMES FERGUSON: They are provisioned for 60 days from Florence They ought to have 3 tons of flour
>
> PREST YOUNG: They want 5 tons of flour[19]

Behind Willie was the Martin company, which, it was estimated, would need six tons of flour. Young's instructions were to deposit flour and feed at several stations on the trail east—"take one load of provisions to Weber [River], another to [Fort] Bridger, and so distribute it along."

By the close of the meeting, plans had been made to collect from the bishops eleven tons of flour plus blankets and shoes. No supplies were discussed for the Hunt and Hodgett wagon companies, who were not in as great danger as the handcart trains. Later Young would write to Amasa Lyman in England: "A portion of our immigration is very late. . . . It was a great mistake to start them so late."[20] But the tone of this meeting, while businesslike and practical, was not one of alarm and no recriminations were made.

In the morning President Young convened the general church conference. His tone was no longer merely businesslike—whether out of savvy rhetoric, or because he had worried overnight, or because, as he told the people, urgency was the dictation of the Holy Ghost to him:

> I shall call upon the Bishops this day. I shall not wait until tomorrow, nor until the next day, for 60 good mule teams and 12 or 15 wagons. I do not want to send oxen. I want good horses and mules. They are in this Territory, and we must have them. Also 12 tons of flour and 40 good teamsters, besides those that drive the teams.[21]

Later in the meeting President Young repeated his call for teams and wagons. "If the teams are not voluntarily furnished, there are plenty of good ones in the street and I shall call upon Brother J. C. Little, the marshal, to furnish them."

One brother was impressed that the president was in earnest; he seemed moved by a spirit that would admit of no delay.[22] Not only did Young want "60 or 65 good spans of mules or horses," but he wanted them "with harness, whipple-trees, neck-yokes, stretchers, lead chains, etc." and he wanted donors and volunteer teamsters to come straightway to the pulpit.[23] From the women he wanted blankets, stockings, shoes, "clothing of almost any description." This second call must have been effective. A thirty-nine-year-old housewife in the congregation observed that "the sisters stripped off their Peticoats, stockings and everything they could spare, right there in the Tabernacle" to be packed with foodstuffs into wagons, which by the following morning numbered sixteen.[24]

Among the first to come to the pulpit were the missionaries who had just returned: Joseph A. Young, Chauncey Webb, George D. Grant, William H. Kimball, and Cyrus Wheelock. Some men were "volunteered," such as Daniel W. Jones, who owned a saddle horse and was an experienced scout. (After the meeting Daniel Wells took Jones by the arm and announced, "You are a good hand for the trip; get ready." Minutes later Presiding Bishop Edward Hunter told him the same thing. Later President Young's counselor Jedediah Grant told Jones, "I want you on this trip." According to Jones, "I began to think it time to decide."[25]) Others to respond were unattached young men like Harvey H. Cluff, whose brother was with one of the handcart companies: "Being in Salt Lake City and of an ambitous turn of mind, I volunteers to go."[26]

Then there was Ephraim K. Hanks, attached, but in a frontier sort of way which left him free to spend "considerable of my time fishing in Utah

Lake." On the way back from one fishing trip, "an ordinary sized man" appeared to him in the night and revealed that the handcart people were in trouble; would he go to help them? Yes, he would go if he were called. Two days later, when a general call was made at conference, Hanks went forward at once. The following morning "I was wending my way eastward over the mountains with a light wagon all alone."[27]

But the cores of the rescue parties were drawn from the local militias. Each valley and community kept a band of soldiers armed and trained to ride against the Indians whenever necessary—which, since 1850, had been frequently.[28] The Salt Lake City cavalry, commanded by George D. Grant and Robert T. Burton, was a part of the new Nauvoo Legion. Every prominent man in town was an officer, even if his status was only honorary. Many of the active senior officers had been to boot camp in the Mormon militias in Missouri and Nauvoo, and at least four who joined rescue parties had marched together in the "Mormon Battalion." Two Battalion comrades, Edward Martin and Daniel Tyler, were leading the handcart company in most danger.

A younger, select corps attached to the various militias was called Minute Men. These were sons and nephews born, wrote Andrew Jenson, during the early traumatic years of Mormonism when their parents were suffering most.[29] They were rugged and feared nothing, having crossed plains, tended herds, logged, plowed and irrigated, carried mail, and fought Indians (or, because Brigham Young was president, not fought Indians, as one complained).[30] Their job, of which they were manfully proud, was to keep their powder dry and be ready to ride at a minute's notice. They shared a spirit of camaraderie, most having been hardened by Indian tussles in Tooele and Sevier counties in 1853, although some were but seventeen and eighteen and had little real soldiering experience. When Brigham Young and Heber Kimball needed rescuers, they looked to "the boys," who included some of their own sons.[31]

Those volunteers who could ride within twenty-four hours were instructed to rendezvous "a short day's drive" east of Salt Lake between Little and Big mountains. The evening of 6 October the boys and their leaders met in Brigham Young's office to receive priesthood blessings and instructions for their mission. Advice could be but tentative, for the facts were sketchy as to how far the party might have to travel and how much relief would be necessary. All hoped for the best and prepared for considerably worse.[32]

In the meantime and throughout the night, bishops scurried to collect provisions from their ward members. This effort would continue for three months, as wagons rolled out of the city to relieve not only the handcart pioneers but the rescue teams, who would be on the trail weeks longer than expected. The Thirteenth Ward, where many of the rescuers and church leaders lived, sent its "block teachers" out to get subscriptions. One teacher reported as follows:

Amount of Provisions and Team Promised by the following named persons on the Northeast Corner block of 13th Ward:

Alfred Best	nothing	Daniel Mackintosh	10 lbs Flour
Wm Capener	Do	John H. Rumell	10 lbs
Solomon Roserter	Do	Hosea Stout	10 lbs
Bro Lea	Do	J. C. Little	25 lbs---
Elias Williams	Do		55 lbs

Sister Major not at Home
William Camp 2 Yoke of Cattle and Driver

John H. Rumell Clk[33]

The early October effort to collect supplies was too hurried to permit careful record keeping. However, a ledger was sent with Robert T. Burton, clerk of the relief party, in which he was to record disbursements to the immigrants. Remarkably, Burton maintained detailed entries from 31 October, when the relief party first met Martin's company on Greasewood Creek, through 19 November, when exhaustion may have overtaken him.[34] Later the church or tithing store seems to have become the clearinghouse for the supply effort. Before being packed into wagons, goods brought in by bishops and private citizens were itemized in store journals. By 15 December the Tithing Office account entitled "Emigration East" would total $1,294, largely in foodstuffs.[35] Considering that flour was valued at $6 per hundredweight, peaches at 18 cents per bushel, women's dresses at $1 to $2 each, and boots at $6 a pair, this represented a substantial donation, though slightly less than the amount requested.

Brigham Young's call for assistance was so successful that all during the day of 7 October teamsters and wagons left Salt Lake City in apparent eagerness to be the first to reach the handcart companies. Those who had heard the orders to rendezvous camped that night in two large groups, one near Little Mountain and the main party ten miles further east at the foot of Big Mountain. Jones, cook for the Big Mountain camp, was very pleased with the company. "A better outfit and one more adapted to the work before us I do not think could have possibly been selected if a week had been spent in fitting up. . . . We had good teams and provisions in abundance."

By late evening, when everyone had drifted in, the Big Mountain party numbered twenty-seven. They were a seasoned group for the most part, two-thirds of them twenty-six years or older. They included first-rate scouts such as Charles Decker, who had already crossed the mountains forty-nine times (Decker and Ephraim Hanks were partners in a mail company), and Abel Garr with his two sons, who oversaw the church herds on Antelope Island.[36] Some were considered rather rough company by local citizens—they drank, swore, and occasionally wrestled in the streets—but in the rescue effort they would justify their place in pioneer Utah society.[37] Commitment helped them say goodbye to tearful parents and wives. Asa Hawley's wife asked plaintively, "Asa, when will you be back?" He answered, "I do not know, but, little girl, rest assured that when I have performed my duty I will gladly return to my little wife."[38] "All in all," wrote Daniel Jones, "those going were alive to the work and were of the best material possible for the occasion."

The company organized, elected Captain George D. Grant and Robert T. Burton as their leaders, and started east the morning of 8 October.[39] They found the roads hard and dry; this enabled them to travel hard, not stopping until late evening to camp and pasture the animals. From the very beginning they felt anxiety for the immigrants. The first night beneath Big Mountain they had experienced a light snowstorm. If they, "strong men with good outfits," found the nights severe, what, they wondered, must be the plight of old men, toddlers, and women trying to pull handcarts?[40]

By Franklin Richards's estimate, Captain Willie would be found in the vicinity of Green River, two days beyond Fort Bridger or about 130 miles east of Salt Lake Valley. Grant's party reached the fort on October 12. Here they purchased some beef and left feed and flour for the return trip, but they were unable to learn anything about the handcart companies. The next day they came upon some of the front-running teams who had "got tired of waiting" and were returning, having decided that the immigrants must have wintered somewhere. With these teams, and others arriving daily from the rear, the relief train soon numbered forty wagons.[41]

On the same day the party met Captain Smoot in advance of the church train and discovered that he too knew nothing of the handcart companies' whereabouts. Grant was dismayed. He immediately sent a four-man express on fast horses and with a light wagon to hunt down the companies and let them know help was on the way. Captain Smoot sent his train, reinforced by fresh teams and eighteen teamsters, on to the valley, while he himself turned east with the express. The main rescue party, headed by Captain Grant, continued on more slowly toward the missing companies. On 15 October "our hearts began to ache when we reached Green River and yet no word of them."[42]

For three days storms passed "to the right and left" of Grant's party. But on 17 October, when they crossed the continental divide at South Pass, a storm met them head-on. From that time on they encountered increasingly cold days and bitter nights. When an animal was killed to take to the immigrants, there was no need to salt the beef—it froze during quartering and stayed frozen. It now seemed necessary to leave more teams behind so that they might be able to assist when the immigrant parties came through. Therefore, Reddick Allred was stationed on the Sweetwater River with flour, cattle, eleven guards, and four wagons. He would soon be joined from the west by three additional wagons and six men.

Had anyone in the relief party foreseen the condition of either of the handcart companies, they would have gathered all the stores and teams at Fort Bridger, Green River, and South Pass and traveled day and night until their animals broke. It was just as well they did not know, for the relief effort would already require more strength and supplies than they carried. Their pace would already stretch every man to the breaking point. Captain Burton had seen rigorous action in Tooele County while chasing Indian marauders; his company had been caught in summer with no water and in winter with no tents, bedding, or warm coats. But of the present campaign he would later state, "This was the hardest trip of my life."[43]

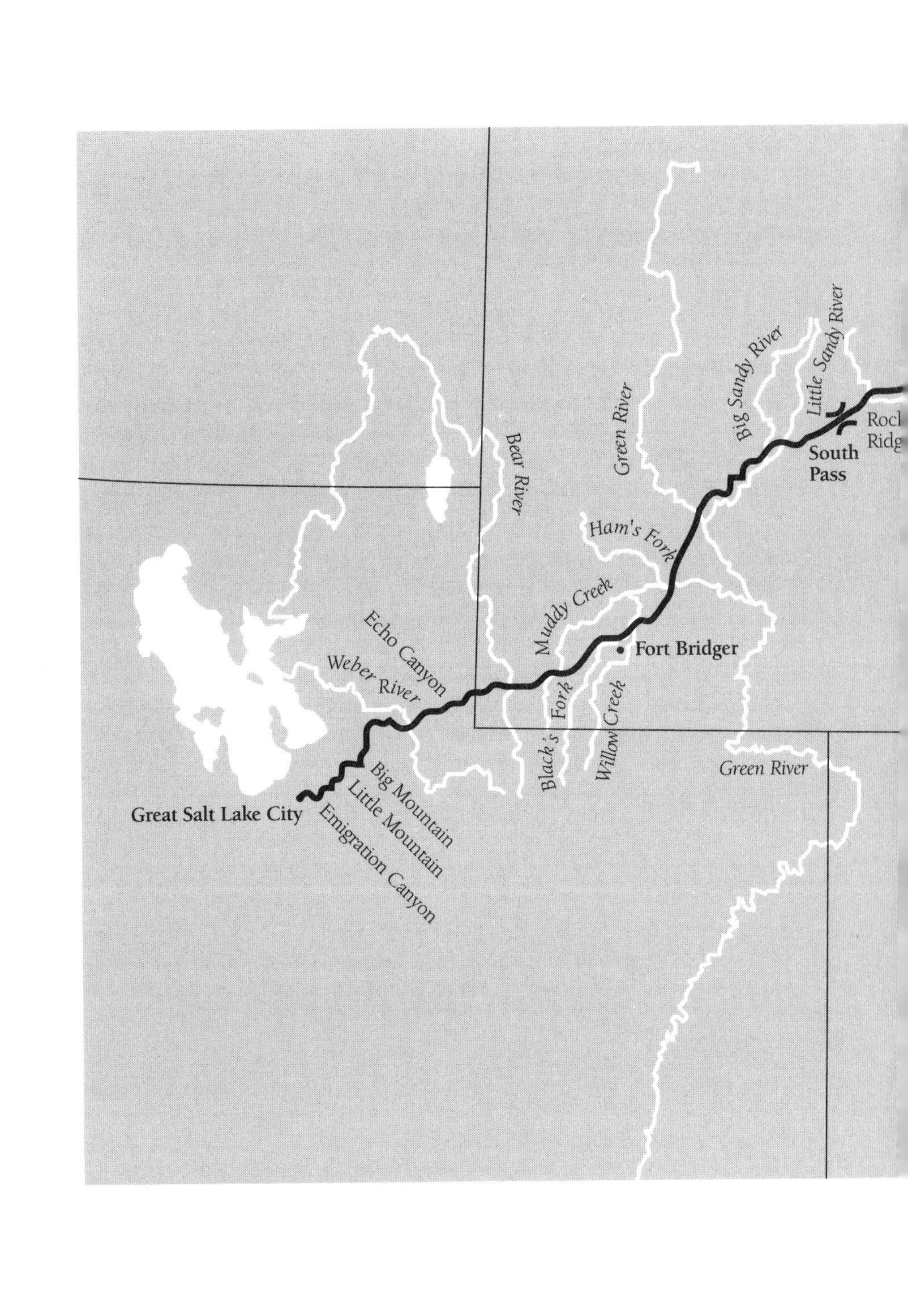
Little Sandy River
Big Sandy River
Green River
Bear River
Rock
Ridg
South Pass
Ham's Fork
Muddy Creek
Echo Canyon
Weber River
Fort Bridger
Black's Fork
Willow Creek
Green River
Big Mountain
Little Mountain
Emigration Canyon
Great Salt Lake City

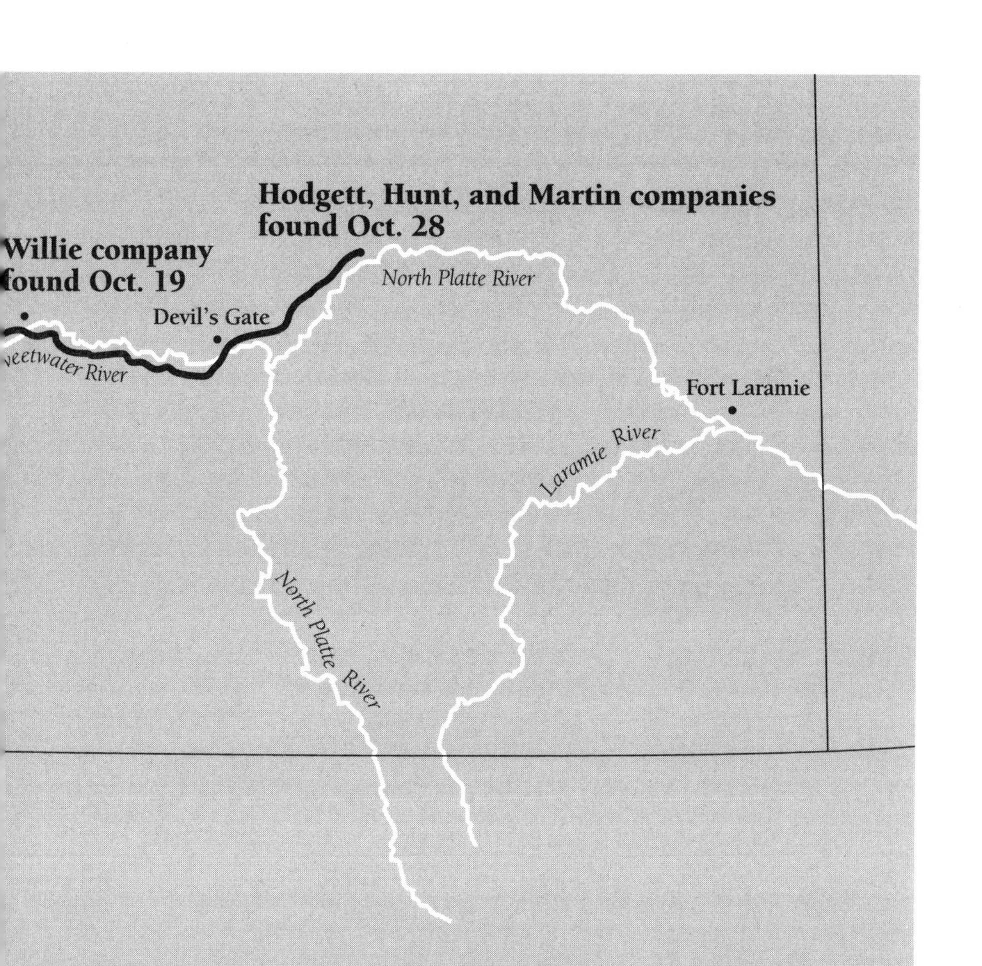

Route of the Rescue Parties

Fort Bridger. Photograph courtesy of Utah State Historical Society.

Rescue of the Willie Company

About 18 October Grant stopped his party at Willow Creek, unwilling to go further through the snow until he was sure of a destination. At that time Willie's company was stranded a day's drive away, near the Fifth Crossing of the Sweetwater. Their condition was extreme. The storm which had delayed the rescue party on South Pass had stopped the immigrants in their tracks. The young, the old, and the weak had already been dying quietly; even "strong men, who were secretly giving their portion [of rations] to their families," began to die in the night.[44] Immigrant Jens Nielson, captain of a tent which in Iowa had housed five men and fifteen women and children, was now the only man left of the five.[45] Sarah James, eighteen years old, was "cold all the time. There was either rain or snow or wind. Even when you wrapped up in a blanket your teeth chattered."[46] Mary Hurren (Wight) ate her three daily ounces of flour cooked as a gruel in a communal family pot. For hunger pains, she and her sister and brother chewed on pieces of rawhide stripped from the cart wheels. Her shoes were torn at the seams and had holes in the bottoms, and at night there was very little bedding with which to bundle up.[47] Everyone's clothing was in rags; for shoes many tied strips of burlap or canvas about their feet. On the night of the eighteenth Emma and Sarah James's father told his children that the rations were gone.

Laleta Dixon's history quotes from Sarah:

> Father was white and drawn. . . . Mother was worried about him, for he was getting weaker all the time and seemed to feel that there was no use in all the struggle. Mother had taken as much of the load off his shoulders as she could in pulling the cart. We girls and Reuben did most of the work so that father could rest a lot. Mother didn't have much to say, and I wondered if she remembered that council meeting in Camp Iowa and wished that we had taken the advice of more experienced people.[48]

Captain Willie ordered several animals butchered for what meat could be gotten from their bones, and then for twenty-four hours the company had nothing to eat. The children chewed on bark and tree leaves and rawhide from the boots of those who had died (by now numbering sixty-six). The next morning twelve bodies were found frozen in their tents. Willie started by mule with one companion, Brother Elder, to try to find the aid they expected from the valley.[49]

At noon, apparently on the same day Willie left, Grant's express team found the immigrant camp. Some inhabitants were too numbed to understand that relief was at hand. But to others fully conscious, this one little light wagon with enough pounds of flour to ward off starvation for a day seemed "like a thunderbolt out of the clear sky," its drivers like "messengers from the courts of glory."[50] Actually all the express had to offer was hope and a miserably inadequate offering of flour, for they had to press on toward Captain Martin's camp. But because of this visit, the Willie camp somehow roused themselves and renewed their journey, going sixteen miles that day in spite of "weary teams and hungry people." At their campground the next morning they buried nine people.[51]

In the meantime Willie and Elder found the rescue camp on Willow Creek after two days' struggling through the blizzard and praying they would not lose the road. Grant had moved his party into a hollow some distance off the road among trees near the river. Had it not been for young Harvey Cluff, who had marched through "the northern blast" to post a sign on the main road marking their detour, Willie would have passed by the rescue camp. As it was, Cluff had barely returned and night had fallen when the two icicled men rode into the hollow. Cluff was proud of his extra exertion: "I have always regarded this act of mine as the means of their salvation. And why not? An act of that importance is worthy of record."[52]

One can imagine Grant's relief on meeting Willie at last. His men broke camp long before daylight and drove twenty-five miles (the average day's travel for a wagon train in seasonal weather was sixteen or seventeen miles) toward the Sweetwater camp where Willie had left his company. By now the immigrants had camped again in sagebrush near the river bottoms. From a mile away, they looked like an Eskimo village—"the snow being a foot deep and paths having been made from tent to tent."[53] The rescuers expected to find the people cold and hungry, but what they saw "would stir the feelings of the hardest heart."[54] Babies and children were crying from hunger; parents were gaunt and apathetic. Some were obviously dying, and the limbs of others had frozen black and were rotting.

Even so, some of the people were able to rejoice. Little Mary Hurren "could hear the squeaking of the wagons as they came through the snow before I could see them. Tears streamed down the cheeks of the men and the children danced for joy."[55] Sarah James and her family were "too tired to move, so we huddled in our covers, close to each other for warmth. It was snowing, and we were so tired. Suddenly we heard a shout, and through the swirling snow we saw men, wagons and mules coming toward us. Slowly we realized that help had come."[56]

The rescuers struck out immediately "and dragged up a lot of wood."[57] Flour, potatoes, onions, bedding, and socks were distributed, although there was not much bounty in proportion to the number of needy. Men were cautioned not to deal out provisions too liberally, as gorging with food would not be healthy after sustained deprival.[58] "Soon there was an improvement in camp," wrote Daniel Jones, although "many poor, faithful people had gone too far—had passed beyond the power to recruit." (Nine more died that night.) Nevertheless, as others gathered about the huge fire to enjoy "delicious morsels of food, we came alive enough to thank the Lord for his mercy to us."[59]

During the entire rescue effort resources would prove sufficient to keep the hardy alive but never enough to significantly reduce the suffering. For one thing, much had to be forwarded to the Martin company, still three weeks behind. For another, rescuers assigned to these 440 pilgrims numbered fewer than twenty. But at least there were rations again—a pound of flour per day per person—and William Kimball with six well-loaded wagons and teams to take the burden of labor. The very sick could ride part of the way and the stronger could sleep in wagons at night, with bedding enough not to keep them warm but to keep the flesh from crystalizing. Each morning three or four people would still be found dead in their blankets. Sarah James remembered the rest of the journey as being "terrible with the cold and snow, but we did have food and some hope of getting to Zion."[60]

It must have taken superhuman will for the company to start again, but start they did. On the second day under William Kimball's command occurred the most terrible ordeal of the entire journey: a five-mile ascent of Rocky Ridge during a howling snowstorm. The northwest wind stung and snow lay knee-deep across the trail. There were not enough able men to help those who needed help. Sarah James recalled being "dizzy and sleepy a lot of the time" so that she did not always know what was happening, but several incidents penetrated her consciousness. When she and her sister could pull no longer (her mother was in a wagon by now), one of the new captains from the valley would come up and pull the cart for a time. She watched a man just ahead of her lay down his shafts and start to cry.

> We all wanted to cry with him. One of the captains . . . came up to him and just slapped him in the face. That made the man so mad that he jumped right up and started to run with his cart. I remember that it was a mean way to treat the poor fellow but [I recognize] now that it saved his life.[61]

Finally, she was aware of the forty-foot-wide crossing of the icy Sweetwater River and how, all through the night, people and carts were being brought into camp on the other side. Two to be carried in about 5 A.M. were her father, who had sat down to rest beside the trail and never gotten up, and her little brother Reuben, whose skin was frozen but who was still alive.

> I can see my mother's face as she sat looking at the partly conscious Reuben. Her eyes looked so dead that I was afraid. She didn't sit for long, for she was never one to cry. She put her invalid son in the cart with her baby, and we joined the train.[62]

Before the train started out the morning of 24 October, thirteen bodies were laid in a shallow grave.

On 25 October the Willie company approached South Pass, where they met Reddick Allred with his seven fresh teams and provision wagons. Now more immigrants could ride and carts could be discarded. At Green River ten wagons came to their aid from Fort Supply, a Mormon station near Fort Bridger. Beyond the pass the weather grew milder and teams were constantly coming from the valley with food and clothing, although most had to be forwarded to the Martin, Hunt, and Hodgett companies.

At last, on 2 November, the Willie company was able to leave the last of the handcarts at Fort Bridger. Outside Fort Bridger they met Ephraim Hanks coming east, who reported that "plenty of teams were ahead."[63] William Woodward wrote:

> From the time we left Fort Bridger about fifty wagons were assisting us on to the valley. The name of all the Brethren that came to our assistance I am not acquainted with. But suffice to say the most of them were on hand and kind to the sick and feeble, although some few were very attentive to the fair sex.[64]

The company was now traveling swiftly enough to overtake the Gilbert & Gerrish merchant train. Nine ambulance wagons arrived in the valley on 3 November. That same day William Kimball sped ahead of the main company to inform President Young of "the state of things generally on the plains." But Young had already started several wagons loaded with corn and hay to relieve the teams of rescuers.[65] On 4 November Young wrote to Amasa Lyman, mission president in England:

> Those who are yet out, we hope and expect at this time are all this side of the South Pass but do not certainly know. Large quantities of provisions, clothing, and a great number of teams and wagons have gone back to assist them.[66]

Jedediah Grant was able to report that Sunday to a tabernacle audience that two hundred rescue wagons were already on the road.[67] Non-Mormon mountaineers who sometimes wintered in the Fort Bridger area must have been astonished at the activity. Each day one or more parties stumbling west would cross paths with one or more rescue outfits hurrying east. There were express horsemen traveling between Fort Bridger and the valley, expresses communicating between the companies on the road, expresses between Fort Supply and Fort Bridger. There were the ambulance wagons, which drove from before sunrise to far past sunset each day, trying to get their human cargo to the valley to receive medical aid. There were the Smoot teams turned east to the aid of Willie and Martin. And there were little posts of men and provisions stashed by Captain Grant during his rush from the city. All this took place *before* it was learned at church headquarters that Martin's company had yet to reach South Pass. Thereafter, the rescue effort escalated.

By 6 November Willie's company was halfway down Echo Canyon, only eighty miles from Salt Lake City. Fewer of their numbers were dying; it seemed that those who survived the ordeal on Rocky Ridge had endurance to withstand the rest of the journey. On 8 November, when the company crossed Big Mountain, the Wasatch Range was receiving its first major snowstorms; snow lay four feet deep on Big Mountain and one foot deep on Little Mountain, giving "the poor thinclad Hand Cart Co. a very cold hard time," observed Hosea Stout.[68] But William Kimball, having hurried ahead to the city, now returned with provisions and fresh teams to pack down the roads and speed the crossing. On 9 November the company entered the valley "where the sun was shining brightly. F. D. and S. W. Richards came to meet us near the mouth of the canyon. We formed according to our hundreds and rolled into the city."[69]

Devil's Gate. Photograph courtesy of Utah State Historical Society.

Rescue of the Martin Company

On 21 October, after leaving William Kimball and six teams to assist the Willie company, George Grant's rescue squad numbered eight wagons and about twice that many teamsters and horsemen. The party proceeded east along the Sweetwater, still traveling blindly with no knowledge about the location or condition of the remaining handcart and oxen companies. Perhaps the immigrants had indeed returned to Fort Laramie to winter. In any case, they must be found; and the task promised to be hard. The rescuers had already been on the road four weeks and were running short of feed themselves.

An express team of Joseph Young, Abel Garr, and Cyrus Wheelock had already been sent ahead to locate and cheer the lagging companies. They had been instructed to stop at Devil's Gate, it being considered unsafe for them to travel further alone. No one believed such instructions would be necessary, however, for, according to Franklin Richards's timetable, Martin's company should be at Devil's Gate or beyond by now. More than likely the October blizzard had stalled them there. If so, they had probably weathered the storm without the terrible suffering of Willie's people. Devil's Gate was distinguished by a little log stockade, three or four cabins, and a cove across the river where year-round forage and firewood were available in the form of cedar and scrubby pine.

But the nearer Grant and his men approached Devil's Gate with no word from the express, the more alarmed they became. Their worst fears had been exceeded by the actual condition of the Willie company. They did not really think Martin would have wintered his huge company on the plains. As Daniel Jones recalled,

> The Elders who had just returned from England having many dear friends with these companies, suffered great anxiety, some of them feeling more or less the responsibility resting upon them for allowing these people to start so late in the season across the plains.[70]

They were disappointed but not surprised to find the express waiting at the four-year-old stockade. Grant encamped his party in the cove and no doubt pondered what to do next. He decided to send out a second express with new instructions to "not return until [the immigrants] were found." Young and Garr, this time with Daniel Jones, again set out "with picked saddle horses and a pack mule."[71] The main group remained at Devil's Gate. "Not having much feed for our horses," Grant reasoned, "they were running down very fast, and not hearing from the companies, I did not know but what they had taken up quarters for the winter."[72]

The express team rode hard one full day. Next day their start was delayed until noon after their horses wandered off overnight with a herd of buffalo. Frustrated, the men rode "at full gallop wherever the road would permit" until, at midafternoon, Joseph Young spied "a white man's shoe track in the road." He called to the others. "We put our animals to the utmost speed and soon came in sight of the camp," reported Jones.[73]

Martin's camp, with Hodgett's wagon train "in camp hard by," was situated at Red Bluffs, sixty-five miles east of Devil's Gate. This was where the blizzard had stranded them for nine days—halfway between the last crossing of the North Platte and the first camp on the Sweetwater. To first appearances they were in no worse condition than the Willie company had been. The fifty-six who had already died were "mostly old people," Joseph Young wanted to believe.[74] But the frightening fact was that the men—middle-aged husbands and fathers who had borne the load—were now failing, and more quickly than their wives and children.[75] Many in the camp were demoralized. They "seemed to be devoid of natural affection and would let their family members die off merely for the sake of getting their few mouthfuls of food or perhaps an old blanket that covered them."[76] Many of them, Captain Grant later commented, were "like children and do not help themselves much more, nor realize what is before them."[77] It was said that several young women, while crossing the Platte during the blizzard, had succumbed to a "dementia" from which they would not recover until well after their arrival in Salt Lake City.[78]

The expressmen were aghast at the shattered spirit of some of the men. This aspect of the ordeal may have taken a greater toll than physical deprival. John Chislett of the Willie company observed:

> Men whom I had known all the way from Liverpool, who had been true as steel in every stage of our journey, who in their homes in England and Scotland had never known want; men who by honest labour had sustained

themselves and their families, and saved enough to cross the Atlantic and transverse the United States, whose hearts were cast in too great a mould to descend to a mean act or brook dishonour; such men as these came to me and begged bread. I felt humbled to the dust for my race and nation, and I hardly know which feeling was strongest at that time, pity for our condition, or malediction on the fates that so humbled the proud Anglo-Saxon nature.[79]

This reaction was not merely broken pride but shock resulting from physical and emotional trauma, and the remarkable thing is not that the relief teams saw so much of it but that a good many immigrants escaped it.

The express carried no flour, only life-saving hope. As the team galloped in, the camp awoke from its stupor; of a sudden all was "animation and bustle."[80] "There was a general rush to shake hands," reported Daniel Jones. "Many declared we were angels from heaven." The three angels walked through camp, trying to encourage people with the news that ten wagons awaited them only three days' distance away. Joseph Young had known many of the immigrants—some he had converted—and what he saw caused him to weep. He called the entire camp out and told them to increase their flour rations to one pound per person and kill the remaining cattle. The express "then started full gallop for John Hunt's camp 15 miles further."[81]

Hunt's wagon company was camped near the winter lodgings of some old trappers. "Their tents were pitched in good shape, wood was plentiful, and no one seemed concerned." To Jones it appeared that "they were on the eve of suffering but as yet had not." Here the express squad, unrecognized and scarcely acknowledged, removed to a spot nearer the river and pitched camp. Daniel Jones wrote:

After a while some one sauntered down our way, thinking probably we were mountaineers. These recognized Brother Young and made a rush for our camp, giving the word; soon we were literally carried in and a special tent was pitched for our use.[82]

Next day the order went out among the Martin, Hodgett, and Hunt camps to pack up and move forward. Brigham Young had sent instructions that their only salvation lay in traveling toward the valley, if only a few miles each day; otherwise they would starve before sufficient help could reach them. Garr and Jones, assigned to Hunt's train, had difficulty starting them. Jones observed:

There was a spirit of apathy among the people, instead of going for their teams at once, several began to quarrel about who should go. This made us

> feel like leaving them to take care of themselves. We saddled up to do so. The clouds were gathering thickly for storm, and just as we were about to start it commenced snowing very hard. The heavens were obscured by clouds, excepting a small place about the shape of the gable end of a house. This opening was in the direction of the valley and the sun seemed to shine through with great brightness. We mounted our mules; Brother Garr, pointing to the bright spot in the heavens, said, "Do you see that hole? You had better all get out of here before that closes up, for it is your opening to the valley. We are going." The people, I believe, took this for a warning and soon started for their cattle.[83]

Some hours later Hunt's company came upon the spectacle of Martin's company struggling up "a long muddy hill." Jones abandoned the wagoneers and went to the aid of the handcart people.

> A condition of distress here met my eyes that I never saw before or since. The train was strung out for three or four miles. There were old men pulling and tugging their carts, sometimes loaded with a sick wife or children—women pulling along sick husbands—little children six to eight years old struggling through the mud and snow. As night came on the mud would freeze on their clothes and feet. There were two of us and hundreds needing help. What could we do? We gathered on to some of the most helpless with our riatas [lariats] tied to the carts, and helped as many as we could into camp on Avenue Hill.[84]

On that day and the next, in spite of these conditions, the handcart company traveled thirty miles, further than either wagon company.

As soon as the immigrants were moving, Young rode back to Devil's Gate to take the news to the main rescue party. "Ah! then there was hurrying to and froe!"[85] Ten men were left to haul wood and ready the stockade to house twelve hundred refugees. Captain Grant and the others, after a day's travel, met Martin's company at Greasewood Creek. Although a welcome sight, the relief was somewhat disappointing because Grant was "out of provisions or nearly so." Heber McBride said later:

> But they had sent a man back to Salt Lake with too [two] horses as an express to let the people know our situation. The 10 wagons relieved us of some of our load by taking the sick into their wagons and a few other things such as tents and cooking things. There was 2 men to each wagon and as they were hearty and strong they took upon themselves to [do] all the work about the Camp and the Captens of companies had no more to say. . . . The men from Salt Lake would clean off the snow and pitch the tents and get

wood for all the families that had lost their Father and then they would help the rest what they could.[86]

Heber McBride's perception of the scantiness of supplies was quite accurate. Robert Burton's account book shows that at Greasewood Creek the following were dispersed: 102 pairs of boots and shoes, 157 pairs of socks and stockings, 30 quilts and comforters, 100 frock coats and jackets of various kinds, 36 hoods, 80 petticoats and bloomers, 27 handkerchiefs, 14 neckties, and 8 pairs of mittens. Still, only one of every two persons in the Martin company now possessed a good coat or a pair of stockings without holes, and each tent shared but one good blanket. In the ensuing days 4,120 pounds of flour and two bushels of onions would be distributed; but as the Hodgett and Hunt companies joined the encampment, provisions would once more have to be rationed. The relief teams brought more hope than relief. Hope lay in additional supplies which were expected daily from the valley.[87]

Grant's men spent a morning "stowing our wagons full of the sick, the children and the infirm" before moving the handcart company from Greasewood Creek to Devil's Gate. Harvey Cluff wrote that "it was a Sunday evening when the handcart veterans pulled into the quarters provided for them. Every nook and corner was taken. Wagons and tents were filled to their utmost capacity."[88] As one group moved into the fort George W. Grant (Captain Grant's son) announced, "You are not going to freeze tonight." He had them stand back. "This night," he declared, "I have the strength of a Grant; I never felt so strong before in my life." According to Patience Loader, "He raised his ax and with one blow knocked in the whole front of the building, took each log and split it in four pieces and gave each family one piece."[89] Cool-headed Dan Jones commented on these gestures: "One crowd cut away the walls of the house they were in for fuel, until half of the roof fell in; fortunately they were all on the protected side and no one was hurt."[90]

On 3 November a council was held to discuss whether to winter where they were or push on to Salt Lake City. If they were to winter, Devil's Gate seemed the only likely site. Besides the log stockade and three or four (less one) nearby cabins, the cove beneath the stone bluffs was relatively protected. Buffalo had been seen; and, if provisions came from the valley as expected, wintering might save lives.

But many people needed medical care. And the logistics of supplying a small city throughout the winter were formidable. The decision was made to try to move as soon as the weather broke. Thereupon, another northern blizzard struck. Space evaporated as the wagon companies straggled in. Snow collected to twelve and eighteen inches and the temperature dropped to eleven degrees below zero. People crowded into the stockade for protection from the gale, but there was not room for a third of them. Many of the immigrant men were so weak that it took them an hour to scrape through the crusted snow (using nothing more than tin plates) to clear a space on which to pitch their tents. "The boys" had to drive the stakes for them into the frozen ground.

Captain Grant "was at a loss to know why no assistance had arrived" from the valley.[91] He determined to send a courier to Salt Lake City describing their plight so that a greater effort would be made. He wrote President Young:

> One company is too small to help much, it is only a drop to a bucket, as it were, in comparison to what is needed. I think that in all one-third of br. Martin's company is able to walk. This you may think is extravagant, but it is nevertheless true.[92]

The couriers again proved to be the tireless Joseph A. Young and Abel Garr. Before riding, Young "put on three or four pairs of woollen socks, a pair of moccasins, and a pair of buffalo hide overshoes with the wool on, and then remarked, 'There, if my feet freeze with those on, they must stay frozen till I get to Salt Lake.' "[93]

Despite the discouraging circumstances, the rescue party remained motivated and in good spirits. Grant himself "never felt so much interest in any mission that I have been sent on, and all the brethren who came out with me feel the same." Heber McBride recalled,

> The men was very kind to us, that is those that came from the Valley, for my Sister and me had nothing to do only try to keep ourselves and Mother and our 2 Brothers a[nd] a little sister from friezing. There was 2 men took a great deal of pains looking after us and cairing for Mother . . . one man's name was Thomas Ricks and the other's was Linton.[94]

Patience Loader remembered the rescuers as especially attentive to the widows:

> During the time we was waiting [for supper] a good brother came to our camp fiar. . . . He asked Mother if she had no husband. She told [him] her husband had died two months ago and he was buried on the plains. He had been standing with his hands behind him, then he handed us a nice piece of beef to cook for our supper. He left us and came back with a beef bone. He

> said, "Here is a bone to make you some supe and don't quarrel over it." We felt surprised that he should think we would ever quarrel over our food. Mother said, "Oh, brother, we never quarrel over having short rations, but we feel very thankful to you for giving us this meat for we had not got any meat, neither did we expect to have any."[95]

No doubt many instances of tenderness and endurance on the part of the rescue company are now obscured by time. Said Daniel Jones, "Some writers have endeavored to make individual heroes of some of our company. I have no remembrance of any one shirking his duty. Each and everyone did all they possibly could."[96]

After several nights in the stockade, the handcart people were told to cross the river and camp three miles away in the cove, where they would have increased shelter. Although they knew the order was for their good, crossing seemed an impossible effort to most of the immigrants. The Sweetwater River, though not more than two feet deep at Devil's Gate, was 90 to 120 feet across. Ice caked the banks, but not sufficiently to hold weight, and ice floes clogged the waters. Patience Loader was feeling weak and sick that morning anyway; when she saw the river "I could not keep my tears back."[97] Perhaps all were remembering the terrible crossing of the Platte, before rescue had come, when so many had had their legs and clothing frozen. One man, upon bringing his handcart to the riverbank, asked sorrowfully,

> "Have we got to go through there?" On being answered yes, he was so much affected that he was completely overcome. That was the last strain. His fortitude and manhood gave way. He exclaimed, "Oh, dear! I can't go through that," and burst into tears. His wife, who was by his side, had the stouter heart of the two at that juncture, and she said soothingly, "Don't cry, Jimmy. I'll pull the cart for you."[98]

It turned out that few of the immigrants pulled their own carts across the stream. Members of the rescue party (Stephen W. Taylor, David P. Kimball, George W. Grant, and C. Allen Huntington are mentioned in one account) spent the day in the water pulling carts and carrying people over. When genteel Patience tried to thank David Kimball, he said, "Oh, damn that. We don't want any of that. You are welcome. We have come to help you." With that persistent attention to social amenities which would be hilarious were it not so incredible, Patience and her mother decided they were all good men, kind-hearted, but "rather rough in their manners." The men were no doubt rough in their manners; but they were also more than chilled and, by the end of the day, exhausted. Andrew Jenson later claimed that some of the young rescuers died at early ages from the effects of their heroism.[99]

The handcart camp remained at Martin's Ravine for five hungry days. Each evening help was expected from the valley, and each evening none came. Grant wanted to pack the entire company in wagons and discard all handcarts, but without reinforcements he could not do this. He must try to get the immigrants moving by their own strength. On 8 November the wind waned, the weather became "some warmer," and the people began hunting up their stock in preparation to move in the morning. Three of the rescuers and seventeen young men from Hodgett's freight train were assigned to remain at Devil's Gate until spring to guard the luggage which had been stowed in the fort. Daniel Jones was appointed their captain, and as much as possible in the way of food supplies was left for their subsistence.[100]

At noon on Sunday, 9 November, the weather fine and warm, the handcart and Hodgett companies moved out. Many handcarts were indeed left behind, but only the very weak were permitted to ride in wagons. Heber McBride was disappointed. "My sister and me would see Mother and Peter and Maggie fixed in the wagon, then Ether Jenetta and me would walk along with the others . . . a great many froze their toes and feet." Some now gave up and died—dropping in their tracks, slumping over their broth by the campfire, their hearts stopping at night in their tents. By journey's end this would be the fate of between 150 and 167 of the Martin company—one-fourth of its members. The typical mortality rate for wagon trains was one in twenty at worst.

On 11 November Martin's group was camped where Bitter Cottonwood Creek met the Sweetwater when someone looked up at a spectre moving toward them through the snow. It was that tenacious loner, Ephraim K. Hanks, who had been doing heaven-knows-what since starting out from Salt Lake ahead of the first teams. His story was that he stopped "about three days" with Reddick Allred at South Pass to wait out a "simply awful storm. . . . [I]n all my travels in the Rocky Mountains both before and afterwards, I have seen no worse." He had then proceeded alone, "being deeply concerned about the possible fate of the immigrants," moving very slowly because of the deep snow. He had passed Young and Garr hurrying east, prayed for and shot a buffalo within fifty yards of one camp—which meat he now imparted to the immigrants—and continued three or more days on "my lonely journey."

I think the sun was about an hour high in the west when I spied something in the distance that looked like a black streak in the snow. As I got near to it, I perceived it moved; then I was satisfied that this was the long looked for hand-cart company.[101]

Hanks's heart almost melted within him at the sight of the immigrants. That evening he went about the camp administering to the sick and dying. One man had been pronounced dead by Daniel Tyler. Hanks said to George Grant, Jr., and William Kimball, "Will you do just as I tell you?" They brought warm water and helped him wash the man from head to foot and then anoint him with consecrated oil according to church ritual. Then they "laid hands on him and commanded him in the name of Jesus Christ to breathe and live." The man sat up and "commenced to sing a hymn." Yet in spite of Hanks's and others' blessings, many immigrants lost limbs: "Many such I washed with water and castile soap, until the frozen parts would fall off, after which I would sever the shreds of flesh from the remaining portions of the limbs with my scissors." In the eyes of many of the sufferers, Hanks became a hero.[102]

Besides buffalo meat, Hanks brought an explanation for why no fresh teams had arrived. Many sent out by Brigham Young had gone as far as Fort Bridger before concluding that the Martin company must have perished in the snow, whereupon the teams had either encamped or turned back. A few had ventured further to South Pass, where they tried to persuade Reddick Allred to return also. Allred and his men had refused.

Grant's party heard all this with astonishment. In his account Daniel Jones declined to reveal the names of the returnees, "for it was always looked upon by the company as cowardly in the extreme." Captain Grant knew that couriers Young and Garr would set the erring teamsters right, but just to be sure he immediately started another express to South Pass. These men rode all night and returned at five A.M. with Allred's four teams and flour.[103] Despite such help the Martin, Hodgett, and Hunt companies were again nearly destitute of provisions.

Mouth of Echo Canyon. Photograph courtesy of Church Archives,
The Church of Jesus Christ of Latter-day Saints.

The Delayed Backup Teams

The cause for delay of the relief teams did not lie with the authorities in Utah. The rescue operation as dictated by them was elaborate and constantly updated. Before Willie's company came in, Brigham Young had received sketchy reports of their condition. On 2 November, Young told an audience in the Tabernacle:

> We can return home and sit down and warm our feet before the fire, and can eat our bread and butter, etc., but my mind is yonder in the snow, where those immigrating Saints are, and my mind has been with them ever since I had the report of their start from Winter Quarters on the 3rd of September. I cannot talk about anything, I cannot go out or come in, but what in every minute or two minutes my mind reverts to them; and the questions—where-abouts are my brethren and sisters who are on the plains, and what is their condition, force themselves upon me and annoy my feelings all the time.[104]

The handcart companies had been the topic at each Sunday meeting, and each Sunday names of volunteers had been taken and the following day teams sent east to buttress Grant's party. At one point Brigham Young himself proposed to go out or to at least "make a short visit to Fort Supply and Fort Bridger." On 13 October, in company with eight men, Young and Heber C. Kimball, with their wives Emmeline F. Young and Mary Helen Kimball, had actually left the city fully intending to meet the immigrants on the road. But two days later a doctor was dispatched to bring in the ailing Brigham Young. All but two of the president's party returned, the two presumably having gone on to Fort Supply for news of the immigrant companies' plight.[105]

It is apparent from the fact that Young and Kimball took their wives that in early October they had had little notion of the difficulties that lay ahead for the handcart companies. But all through the month of October partial appraisals had been received. On the road east Captain Grant's party had encountered a number of immigrant and merchant trains heading for the valley. Through the Texas company, which descended Emigration Canyon on 11 October, Brigham Young had learned the progress of the relief party over Big and Little mountains. On 24 October, Dimick Huntington, a scout, had brought further news about Grant and a report that Smoot's church train had been seen on the Green River. The next day a partner in the firm of Livingston and Kinkead had arrived in town with his merchant train and word from Charles Decker, Grant's chief guide: "The snow was lying deep and there was no feed for the horses and [he] requested that Bro Feramorz Little would send some grain."[106]

Although these reports were infrequent, they had been regular enough to help Young gradually surmise the straits of the companies and forward available relief. Church units in the immediate vicinity of Salt Lake City had been called upon in September—the militia to send men; bishops to send teams, feed, and foodstuffs; ward Relief Societies to send bedding, clothing, and medicines. By October outlying settlements—Tooele, Weber, Centerville, Union, Provo—had become the donors. In November and December, as the immigrants arrived and as they recovered, it would become the turn of the far northern and southern settlements to help find them housing and work.

It was 31 October before G. H. Gibbs had "come up" ahead of the Willie company with the first solid information about that company to be received in Salt Lake City. All that morning "Presidents Young and Grant were counselling about the matter."[107] One result was that during the next several days men such as Bishop Silas Richards of Union had been asked to send more teams east.[108] About this time Goudy Hogan and Franklin Standley, with others from northern settlements, were called "to go out with 4 horses to each wagon to haul provisions to the handcart people." They had been told that it would take ten days but once on the trail they were shuffled east to help Martin's company so that it was forty days before they returned.[109] Several weeks later Asa Hawley and others of the Utah County militia had been sent out.[110]

These were the later teams for whom Captain Grant had waited in exasperation. There was nothing cowardly about them, whatever Grant's men might think. First, without hard facts as to where they were going and what was needed, it was difficult to know how much to risk. Second, the roads in places were nearly impassable by late October, whereas for Grant's party the roads had been quite reliable. These later rescuers were in as much danger of frozen limbs and starvation if they pushed too far too fast as were the immigrants. Crossing Big Mountain, Asa Hawley and others had encountered snow "up to the tops of our wagon bows." It reminded him of Bonaparte crossing the Alps. "We ploughed our way through and went on." Some of his companions had their feet badly frozen and had to be left on the Weber, Bear, and Yellow rivers to recuperate. But on the way back, wrote Hawley, these boys made it all up—they were fresh and strong and "took hold with a will, which relieved us very much."

The hardier teams had reached Fort Bridger. There they congregated with other wagons originally sent out to help Willie but who had been commandeered into heading east toward Martin's camp. At Fort Bridger there had been no word on what to do next. C. N. Spencer and John Van Cott continued one day beyond the fort but did not venture a crossing of the Sweetwater, concluding that the Martin company must have either wintered or perished along with Captain Grant's relief teams. They were the ones who had started back toward the valley, causing "all the teams which had gone on the road to help them[,] in all 77 teams which had arrived at Bridger," to return with them.[111]

Fortunately someone at the fort had thought to send a courier to apprise Brigham Young of the relief teams' return. This courier reached the valley 11 November—two days after the Willie company. According to Hosea Stout,

> [T]his news was very unexpected. . . . Immediately upon receiving the news the president sent W. H. Kimball, Joseph Simmons James Furgerson & myself as an express to go and turn the teams East again and for us to find where the Hand cart company was.

William Kimball had been in Salt Lake City but one full day. Yet he and Stout left at sunset and traveled five miles before camping for the night. They found Van Cott the next morning just over Big Mountain; Spencer had already "gone home in the night." Stout continued:

> Van Cott justified himself for returning and abandoning the Hand Cart Company as he could get no information of them and had concluded they had returned to the states, or Stopt at Larimie, been killed by the Indians or other wise gone to the devil and for him to have gone further was only to loose his team and starve to death himself & do no good after all and as for G. D. Grant and those with him who had gone to meet them they had probably stoped at Ft. Larimie. So on these vague conclusions he had not only turned back but had caused all the rest of the teams to return and thus leave the poor suffering Hand carters to their fate.[112]

Such self-consideration was not acceptable in pioneer Mormon society. William Kimball "repremanded him severely" before presenting Van Cott with a letter from President Brigham Young, whereupon Van Cott turned his wagons right around and started east again. No doubt Spencer, when caught up with by Brigham Young, got a similar tongue-lashing.

Precious days, perhaps weeks, had been lost. Apparently Stout, Kimball, Van Cott, and the teams they met traveled night and day, for they arrived at Fort Bridger at noon on 15 November, after only three and a half days' backtracking. On the Weber they had encountered Young and Garr and now knew that the handcart company was encamped at Devil's Gate. A beef was killed at the fort to be sent by quick, light mule team to the immigrants. During the next two days they overtook ox teams which had already been redirected by Young and Garr. Some they advised to camp so as to be rested when the immigrant company was brought through. Others joined their caravan.

On the sixteenth Stout's party "camped at Big Sandy with several teams from Centerville and a large number of oxen from Fort Supply, all who were hurrying on to meet & relieve the H. Carts." They found a load of provisions strongly guarded on the Little Sandy and another team at Pacific Creek near South Pass. "Our train now began to look quite large," reported Stout, "being some 30 wagons."[113]

At last, on 18 November, Stout's men met an advance team from Grant's company "at the Station on the Sweat Water," who informed them that the immigrants would arrive there during the night.

> Several teams were dispatched to meet them and help them in. Soon they began to come in, some in wagons, some on horses, some on foot, while some hed to be lead or carried on the backs of men.[114]

Rescuers and immigrants alike were "filled with delight" to meet Kimball and Stout. This new relief came just in time for those on foot who could not have endured the next day's ordeal of scaling Rocky Ridge. Now there were enough wagons for all the people to ride, though a durable few elected to walk, saying they kept warmer that way. Hosea Stout gives a latecomer's picture of the Martin company in camp the night after crossing South Pass.

> Some were merry and cheerfull some dull and stupid some sick some frosted & some lazy and mean but all seemed to be elated more or less with the idea of speedily arriving in the Valley.[115]

Stout's party, buttressed by four-horse teams sent out after his departure from the valley, gave the relief effort all the labor necessary to speed the immigrants in. The Martin company now traveled twenty-five to thirty miles a day. Light ambulance wagons, with the freshest men and teams, hurried ahead of the train carrying critically ill patients, mostly small children. Of the remainder of the company, two or three more died each day while the survivors, except for a sturdy few, could no longer act for themselves. The rescuers thus became orderlies, with a hospital-like routine. Asa Hawley wrote:

> We had given up our wagons to them [the immigrants]. After arranging their beds as well as we could when bedtime had come, we would carry them to our wagons. After seeing them to bed, we would close the wagon covers thus shutting out all the cold possible. Thus we would leave them for the night. Then shoveling away the snow we would lay our scanty blankets down for a little rest, then up in the morning a long time before daylight we would build a big fire and prepare breakfast. When all was about ready we would arouse our passengers, again repeating that which we had done the day before. When we were all seated we would again pass them their food. Breakfast over, all was now a hurry and bustle to be off. . . . We again loaded them into our wagons and traveled on. This was repeated night and morning all the way.[116]

A letter written by Joseph Simmons, one of Stout's party, portrays the handcart company in camp on one evening:

> Handcart Company Camp
> on Muddy 12 miles from Bridger
> Nov 24th 1856
>
> Bro Horace [K. Whitney]
>
> As some of our Company start for the City tomorrow morning I improve the opportunity by writing you a few scrawls. I am sitting not on the stile Mary [type of desk?] but on a sack of oats with the paper on my knee, by the side of a blazing camp fire, surrounded by some eight hundred persons. One old lady lay dead within twenty feet of me, babies crying, some singing, some praying, &c, &c, but among all this I feel to rejoice for the hand of the Lord has been continually with us. Almost every day angry storms arise very threatning and judging from their appearance one would think that we should be unable to withstand the tempest, but the prayers of the holy men of God are heard. The clouds divide to the right and left, letting the saints pass through in safty. The suffering of the camp from

frozen feet and various other causes I will not attempt to describe. Suffice to say bad, bad, bad. The boys, including your humble servant, are all enjoying the blessing of health, lonely in their feelings, and doing all in their power to make the saints comfortable. We have some seventy waggons, divided into six divisions one captain appointed over each, but all make one camp at night. We intend reaching the valley next Saturday [29 November], but this calculation is founded upon the faith of our heavenly Father being continually with us, staying the storms as in the past, for without the help of high heaven, we should have been snow bound in the mountains long ago. . . . Bro Burton is writing to Bro Brigham and will probably present more fully the situation of the camp. At all events this is all you will get from me this time. Give my love to everybody.

May our heavenly Father bless Bros Brigham, Heber, Jedediah all the saints throughout the world & all honest hearted people.

Your devoted friend J M Simmons

[In margin:] John Whitney says he is all right except frosted toes.[117]

Captain Grant now felt free to leave the entire operation in charge of Robert Burton and, with Stout and three others, head "quite briskly" toward the valley to give a final report to President Young. He knew the immigrant companies were not yet out of danger, for he himself "could scarsely stand alone or keep awake" by the time he reached Fort Bridger. At Bear River he encountered a severe snowstorm and "the coldest and most piercing weather we have had during our journey." His purpose was to report the condition of the three needy companies and get the president's concurrence on the decision to bring in the people but to cache the luggage at Devil's Gate. He also wanted extra teams to tamp the trail ahead of the immigrants.[118]

But President Young was a step ahead of him. By the time Grant reached Big Mountain he found Joseph Young, Brigham Young, Jr., Feramorz Little, and others with ox teams already breaking the road through waist-deep snow. After a conference between Grant and the First Presidency, more teams were sent to Echo Canyon and Bear River. Also, wrote the church clerk, couriers "started for Bridger this morning to instruct the Ox Trains company to remain at Fort Supply & Bridger during the winter."[119] In this way the "church herd" (consisting, complained Harvey Cluff, mostly of privately owned Texas cattle which he was "called" to tend without recompense) would survive until spring.

By 28 November the first of the Martin company were crossing Big Mountain. For a week the weather had held, but now there were four feet of fresh snow with twenty-foot drifts on the north side of the summit.

With help from the tamping teams, by stomping main and side tracks and shoveling through the drifts, Burton's men brought the wagons over. A camp was made in East Canyon, to which men from the valley began to filter. Some of these hurried back to advise Brigham Young on the company's progress. By the time the immigrants camped at the head of Emigration Canyon the following evening, fresh supplies were waiting for them. At noon on Sunday, 30 November, as a congregation exited after services, 104 wagons carrying the Martin company members rolled past the old Tabernacle and halted before the tithing offices where the former Hotel Utah now stands.[120] On 7, 10, and 15 December, some fifty more wagons, carrying 360 members of the independent wagon trains who had rested at Fort Bridger for a time, reached the city. Of these, however, "only a few were suffering from the cold, and their condition far preferable to the last {handcart} companies."[121]

Main Street, Salt Lake City, Utah. Photograph courtesy of Utah State Historical Society.

they were transferred to settlements throughout the territory. Bishops had been alerted in advance to send teams and word of how many immigrants they could provide for. The Cunningham family was "sent to American Fork where my home has been ever since."[133] The Cantwells moved to Cottonwood, Rebecca Pilgrim to Lehi. Said John Oborn, "We lived for a short time in the Fifth Ward, then we moved to Union Fort. We soon went to live by ourselves."[134] Wrote Patience Loader, "Brother William Thorn, bishop of the Seventh Ward, took home with him my mother and my brother Robert, myself, and sisters Maria, Jane, and Sarah. My sister, Tamar, went with Brother Thomas Ricks to Farmington. This is the kind brother that gave us the beef at Devil's Gate."[135] Silas Richards, bishop of Union, was directed by Bishop Hunter to take three orphan sisters to raise. Richards's diary entry gives their names:

> Susannah, age 11, Sarah Ann, age 4½ and Martha age 2; the daughters of Daniel and Susannah Osborn who died on the plains. The children are very poor from suffering with cold and hunger and the want of proper nourishment and clothing. None of them were able to walk across the room without difficulty. Susannah's feet were badly frostbitten, their only clothing being some remnants of summer apparel. They had no shoes or woolen stockings; they were very feeble and required much care and attention.[136]

Mary Ann James was taken "in a day or two" to Springville, where the bishop "had sent for one family of immigrants for whom he was willing to care." There the older James children boarded at different homes in exchange for work while their widowed mother and small children were provided a simple shack. It lacked furniture and a stove, but it was sufficient for the winter.[137] The McBrides stayed for some time at Samuel Ferrin's house in Ogden.

> After resting awhile Mother began to get a little better and then Bishop Chauncy West got a little log house for us to live in, but we had nothing to put into it. There was a fireplace but no furniture of any kind, the snow still deep but people was very kind to us and furnished us with wood and provisions.[138]

In the spring the McBrides moved north onto a farm belonging to Brother Ferrin, and a year later Mrs. McBride and Ferrin, a widower, were married. Ferrin "proved himself to be a kind Father and we all got along firstrate."

Some immigrant women married into the families by which they were employed, often in polygamy. Thisbe Read, for instance, married Ephraim

Hanks as a plural wife. Jane Griffiths remembered staying first in one house three weeks, then being taken in by Brother Mulliner, who had heard that she was to be sent to "the poor house in Provo" and had responded, "Never as long as I have a home." He carried her in a wheelbarrow to his house where she stayed bedridden all winter until, in March, she walked again for the first time. Then she was taken to a friend of her late father. As she wrote:

> He said he would keep me. I went there on my 9th birthday. Mother would carry me as far as she could then put me down in the snow and we would cry a while and then go on. I did not stay there long, they told me I would have to hunt another home. I did not know where to go and I was on the woodpile crying when Brother William Keddington came along and wanted to know what was the matter. When I told him he said, "You come along with me and you shall have a piece of bread as long as I have one to break." He afterward married my stepmother and I had a home with them as long as I wanted one.[139]

In this pioneer territory where personal self-sufficiency was an unquestioned ethic, immigrants were to be given a start only until they could support themselves. No systematic welfare system existed beyond the tithing receipts and allotments by individual bishops. These donations, often in the form of live animals, grain, or other foodstuffs, were slender in 1856 because of the preceding year's drought. Thus the handcart immigrants were dependent on the graciousness of established settlers. The handcart arrivals did not take this assistance for granted. Yet, after all, this was Zion, where they expected no less. Economy governed; immigrants were needed and room was gladly made for them on job sites, in shops, in families. But response to the plight of the Willie and Martin handcart companies was evidence that economy in pioneer Utah was in turn governed by considerations peculiar to the Kingdom: charity to an extent requiring sacrifice, cooperation, and obedience to direction.

Conclusion

The handcart tragedies could have been significantly divisive of the Mormon community. Hardly a family was not connected in some way with one of the immigrants. Only the very young had not crossed the plains themselves and could not identify intensely with the difficulties the Willie and Martin companies had encountered. A community pride and unity had grown out of the sacrifices made in Ohio, Missouri, Illinois, and Winter Quarters; but here was a sacrifice unparalleled in the Mormon experience. The depth to which the episode affected the Mormons can be seen in one result of it, the heightening of the Reformation of 1856–57, which began with Brigham Young saying:

> I should be . . . thankful if the minds of this community could be so impressed and stirred up, so wakened up, that when these poor brethren and sisters who are now on the Plains do arrive they may be able to say of a truth and in very deed, "God be thanked, we have got to Zion." But fearfulness and forebodings of disappointment to them are in my feelings. How far they may be disappointed, I do not know.[140]

There must have been much pondering, in spiritually inclined minds and hearts, as to why such a disaster was allowed to happen and what the causes were. The misunderstanding between the Liverpool and New York offices of the church's Perpetual Emigrating Fund; the rush to build handcarts at Iowa City, even to the point of using improperly cured lumber; the large proportion of old people and children in the companies; the unprecedented early snow and freezing temperatures (conditions unimaginable to immigrants used to the mild winters of England and Denmark); the willingness of the immigrants to trust that all would work out well—all of these were mentioned by those who left a record as causes of the tragedy. Most Utahns and immigrants, like Job, accepted the episode as one of the sacrifices involved in doing God's will and getting to Zion. Those who attributed it to human failure—to bad judgment—and

Brigham Young was one of these—somehow had to resolve their bewilderment and rage. In deploring the suffering of both the immigrants and their rescuers, Brigham Young placed the major blame on the inexperience in "public business" of the immigration officials:

> If they had sent our immigrants in the season that they should have done, you and I could have kept our teams at home; we could have fenced our five and ten acre lots; we could have put in our fall wheat; could have got up wood for ourselves and for the poor that cannot help themselves . . . whereas, now your hands and mine are tied.
>
> This people are this day deprived of thousands of acres of wheat that would have been sowed by this time, had it not been for the misconduct of our immigration affairs this year, and we would have had an early harvest, but now we may have to live on roots and weeds again before we get the wheat.[141]

If to some historians this appears to be scapegoating, to the Mormons it seemed to satisfy the demands of justice, and there was no further official rebuke. Mercy had been satisfied by the rescue effort itself, stories of which soon came to be a dominant theme in the retelling of the 1856 handcart episode. The church, the doctrine of gathering, and—somewhat shakily—the handcart plan were reaffirmed. "Brigham Young," wrote T. B. H. Stenhouse, "did all that man could do to save the remnant and relieve the sufferers. Never in his whole career did he shine so gloriously in the eyes of the people. There was nothing spared that he could contribute or command . . . good teams, flour, beef, potatoes, warm clothing and bedding."[142] As for the "heroes" of the rescue—"the stout hearts and strong hands of the noble fellows who came to our relief"—God had helped them to do it, and all in all the immigrants had "great reasons to thank God for the Saints in Utah."[143] Memories of what was perhaps the worst disaster in the history of western migration have been palliated by what could also be regarded as the most heroic rescue of the Mormon frontier.

Appendix

The following list of forty-eight handcart rescuers and biographical data on them has been compiled from Andrew Jenson, *Latter-day Saints' Biographical Encyclopedia,* 4 vols. (Salt Lake City: Andrew Jenson History Company, 1901–36); Solomon F. Kimball, "Our Pioneer Boys," *The Improvement Era* 11 (July 1908):668–74; Kate B. Carter, ed., *Heart Throbs of the West,* 12 vols. (Salt Lake City: Daughters of Utah Pioneers, 1936–51), and Kate B. Carter, comp., *Treasures of Pioneer History,* 6 vols. (Salt Lake City: Daughters of Utah Pioneers, 1952–57); Autobiography of Asa Hawley, Archives of The Church of Jesus Christ of Latter-day Saints, Salt Lake City (hereafter referred to as Church Archives); Papers of Dan Jones, Church Archives; Diary of Goudy Hogan, Utah State University Library, Logan, Utah; Juanita Brooks, ed., *On the Mormon Frontier: The Diary of Hosea Stout,* 2 vols. (Salt Lake City: Utah State Historical Society, 1964); Autobiography of Harvey Cluff, Church Archives; and family group sheets in the Family History Library, The Church of Jesus Christ of Latter-day Saints, Salt Lake City.

Thomas Murphy Alexander. Born 1831 in Tennessee, son of Randolph Alexander. Parents baptized 1836 and 1837. Thomas named a son David William Patton, which suggests that his family had associated with David Patton in Missouri. By 1839 the family was in Quincy, Illinois, and in Salt Lake City 1848. Thomas had in all sixteen children by his two wives. Died 1905 in Idaho. Age twenty-six at time of the handcart rescue.

Reddick N. Allred. Born 1822 in Tennessee. Served in Mormon Battalion. After 1853 Black Hawk War a colonel in Utah territorial militia. Age thirty-four at time of rescue.

"Tom" Bankhead. Daniel Jones listed "Tom" as only "colored man" in rescue party. Possibly this was Dan, a blacksmith for John Bankhead of Tennessee, who migrated to Utah in 1848 with "Mississippi Saints." Dan and his brother Nathan were dispatched with teams from Salt Lake Valley at least two different years to bring Saints to Utah. Descendants still in Salt Lake Valley.

William K. Broomhead. Born in England in 1833. Moved with family to Nauvoo, Illinois, in 1842 and to the Salt Lake Valley in 1852. Carpenter on the Salt Lake temple. A photo of him in Kimball's *Improvement Era* article places him with Minute Men. Died in Bloomington, Bear Lake Valley, in 1877. Age twenty-three at rescue.

Isaac Bullock. Born 1824 in New Hampshire, son of Benjamin Bullock and Dorothy Kimball. Settled Fort Supply 1855. Married Elect Wood in Provo, 14 December 1856. No reference in his journal to handcart experience, but mentioned by Cluff as one of rescuers. Age thirty-two at rescue.

Robert T. Burton. Born 1821 in Canada, where his parents had emigrated from England. As a teenager was baptized with his family. Served in Nauvoo Legion and Nauvoo Brass Band. When Legion reorganized in Salt Lake City in 1849, he was made bugler in the first company under George D. Grant. Members of this company became heroes in winter 1850 during "Provo River Battle" with the Utes. In 1853 had another hard experience in Tooele County when fighting Indians in winter with no shelter or bedding. Was made militia major and colonel in 1855–56. Age thirty-six at time of rescue.

Harvey H. Cluff. Born 1836 in Ohio, living in Provo by 1856. Served in "Standing Army after Echo Canyon War." His diary gives detailed account of rescue. Twenty years old at the time.

William M. Cowley. According to an account of the rescue in Carter, *Heart Throbs* 1:80, met Emily Wall of the Martin company at Devil's Gate, afterwards married her. Was a "very young printer." No further information found.

"Handsome Cupid." So listed by Dan Jones. No further information found. (No doubt the reader's ancestor.)

Charles F. Decker. Born 1824 in New York. Married 1847 in Florence, Nebraska, to Vilate Young, daughter of Brigham and Miriam Works Young. Traveled Mormon Trail through Rockies forty-nine times prior to rescue, according to George D. Grant. Age thirty-two at rescue.

George Dunn. Sent in November from Union Ward, according to Bishop Silas Richards. No further information found.

Amos Fairbanks. Perhaps the Amos Fairbanks born 1802 in New York. He was living in Indiana in 1842, baptized 1853, endowed 1854, died 1882 in Bountiful. He would have been fifty-four at time of rescue and possibly too old to have participated, but no record could be found of another Amos Fairbanks in Utah at that time.

James Ferguson. Born 1828 in Ireland, served in Mormon Battalion. An actor with Joseph Simmons. Member of Stout's rescue party. Died 1863. Age twenty-eight at rescue.

Abel Garr. Solomon Kimball describes him as "a prominent scout in early days and had charge of Church herds on Antelope Island." Photo in Kimball's article.

David Garr. Presumably one of "the Garr boys," sons of Abel. No other information found.

John Garr. Presumably one of "the Garr boys." No other information found.

Henry Goldsbrough. Possibly the Henry Goldsbrough who had a son Henry, born 1885 in Nephi, the son of Ellen Russell Goldsbrough. No other information found.

George D. Grant. Accompanied Edwin D. Woolley and church herd to California 1853. Was in brigade of Minute Men who protected settlers from the Indians. Age forty at rescue.

George W. Grant. Son of George D. Grant and nephew of Jedediah Grant (second counselor in First Presidency of Mormon church), according to Solomon Kimball, who includes a photo. One of three eighteen-year-olds Kimball credits with carrying the Martin company over the Sweetwater Crossing during a blizzard.

Charles Grey. No information found.

Benjamin Hampton. Born 1837 in Philadelphia, parents from Chester County. A brother (or sister) was born 1849 in St. Louis. Benjamin was endowed 1855, married Adelaide Eugenie Grant 1864, had a daughter who married William S. Godbe. Photo in Kimball article. Age nineteen at rescue.

Ephraim K. Hanks. Born 1827 in Ohio. In Mormon Battalion and 1853 Walker War. Scout and Indian agent. Age twenty-nine at rescue.

Asa Hawley. Born 1835 in Canada, converted with parents in Iowa, came to Utah 1852. In S. S. White's and Sidney Willis's (Lehi Ward) Minute Man companies. Escort to George A. Smith who was, in 1856, commander of Southern Military Department (militia). Age twenty-one at rescue.

Goudy Hogan. Born 1829 in Norway, emigrated 1849. In fall 1856 "I was called on with several others to go out," he says in diary. Age twenty-seven at rescue.

Clark Allen Huntington. Married Rosanna Galloway, had a son born 1859 in Salt Lake. Photo in Kimball article. One of three eighteen-year-olds to carry immigrants over Sweetwater at Devil's Gate.

James Ivie. Probably the Utah militia colonel who was born 1833 in Missouri, fought in 1853 Walker War. Died 1909. Age twenty-three at time of rescue.

Daniel W. Jones. Born in Missouri. In Mormon Battalion. Baptized 1851 after arriving in Salt Lake City. Wrote important account of rescue effort. Died 1915. Age twenty-six at rescue.

David Patton Kimball. Born 1839 in Commerce, Illinois, son of Heber C. and Vilate Kimball. One of three eighteen-year-olds praised by Solomon Kimball. Died 1863.

Heber Parley Kimball. Born 1835 in Ohio, son of Heber C. and Vilate Kimball. Died 1885. Age twenty-one at rescue.

William Henry Kimball. Born 1825 in New York, son of Heber C. and Vilate Kimball. Was returning from mission to England at time of disaster; arrived in Salt Lake City with Franklin D. Richards; left again immediately to find Willie company. Died 1907. Age thirty-one at rescue.

Samuel Linton. Born 1827 in Ireland. Emigrated to Utah in 1853. Age twenty-nine at rescue.

Feramorz Little. Born in New York, nephew of Brigham Young. Was mail partner with Hanks and Decker, enduring great hardships. After rescue took mail trip to Independence, Missouri, arriving early 1857. Age thirty-six at rescue.

John Riggs Murdock. Born 1826 in Ohio. In Mormon Battalion, later with Brigham Young Express and Carrying Company. Age thirty at rescue.

Ira Nebeker. Born 1839 in Illinois. In Burton's company of Minute Men 1855. Died 1905. Age seventeen at rescue.

C. Nowlan. Mentioned by Silas Richards as being sent out in November from Union Ward with supplies to refurbish rescue party. No other information found.

Joel Parrish. Cook mentioned by Dan Jones. No further information found.

Edwin Martin Peck. Born 1828 in Vermont; settled Provo. Died 1903. Age twenty-eight at rescue. No further information found.

Thomas Edwin Ricks. Born 1828 in Kentucky. Explored with Parley P. Pratt. Took Indian mission to Nevada. Photo in Kimball's article. Died 1901. Age twenty-eight at rescue.

Joseph Marcellus Simmons. Born 1824. Son-in-law of Bishop Edwin D. Woolley. During summer and fall 1855 had been at Fort Supply to control Indians and to help immigration. On staff of Colonel George D. Grant; later associated with Burton in Echo Canyon War; colonel by October 1857. Age thirty-two at rescue.

C. N. Spencer. No information found.

Franklin Standley. Born 1831 in Ohio. Migrated with parents to Utah in 1852; settled in Bountiful. Died 1859. Age twenty-five at rescue.

Hosea Stout. Born 1810 in Kentucky. In Missouri militia, Nauvoo Legion in Nauvoo and Salt Lake City. Head of Salt Lake City police and prominent lawyer and diarist. Died 1889. Age forty-six at rescue.

Stephen Wells Taylor. Born 1834 in England. Married (1857) Harriet Seely Young, daughter of Brigham Young. Died 1920. Age twenty-two at rescue.

John Van Cott. Born 1814 in New York. President of Scandinavian Mission 1853–56; had returned with Franklin D. Richards. Died 1883. Age forty-two at rescue.

Chauncey G. Webb. Born 1836 in Kirtland, baptized 1849, married daughter of Seth Taft 1855. Handcart agent in Iowa. Daughter Ann Eliza Webb divorced Brigham Young. Age twenty at rescue.

Cyrus Hubbard Wheelock. Born 1813 in New York. Chaplain in main rescue party. Gifted orator and scout. Died 1894. Age forty-three at rescue.

Brigham Young, Jr. Born 1836 in Kirtland. A Minute Man. Helped tamp roads on Big and Little mountains for hardcart companies. Age twenty at rescue.

Joseph A. Young. Born 1834 in Ohio, son of Brigham Young. Had returned from mission to England (1854–56) with Franklin D. Richards. Died 1875. Age twenty-two at rescue.

Notes

1. Brigham Young to Edmund Ellsworth, 29 September 1855, Archives of The Church of Jesus Christ of Latter-day Saints, Salt Lake City (hereafter referred to as Church Archives). Ellsworth's given name is sometimes spelled Edmond. Original spellings and usages in this quotation have been retained; however, some quotations elsewhere have been altered, for the sake of clarity, to reflect modern spelling and usage.

2. For analysis of this critical period in the Utah economy, see Leonard J. Arrington, *Great Basin Kingdom: An Economic History of the Latter-day Saints, 1830–1900* (Cambridge, Mass.: Harvard University Press, 1958), pp. 131–60.

3. See conference report of Franklin D. Richards delivered in the Bowery, Salt Lake City, 5 October 1856, as published in Brigham Young et al., *Journal of Discourses,* 26 vols. (Liverpool: Latter-day Saints' Book Depot, 1852–86), 4:114–19 (hereafter referred to as *JD*). The evening before this speech Richards had informed the First Presidency: "There were 20,000 over there that were panting to pull a hand cart." See Minutes of Special Meeting in the Historian's Office, 4 October 1856, General Minutes Collection, Church Archives.

4. LeRoy R. Hafen and Ann W. Hafen, *Handcarts to Zion* (Glendale, Calif.: Arthur H. Clark, 1960); and Wallace Stegner, *The Gathering of Zion: The Story of the Mormon Trail* (New York: McGraw-Hill, 1964). We have benefited from edited transcripts of several journals and speeches included in the appendices to the Hafens' book, although all were checked against the original manuscripts.

5. *JD,* 4:117.

6. Hafen and Hafen, *Handcarts,* p. 140.

7. Andrew Jenson, "The Belated Emigration of 1856," *The Contributor* 14 (January 1893):134.

8. In makeup the Willie and Martin companies resembled earlier trains: One-third of the members were children under thirteen, one-third women, one-quarter men, and one percent over sixty-five years. The Martin company was half again as large as other companies; half of its families were headed by widows (much more than half by the time they reached South Pass); and, while there were plenty

of bachelors who might have shouldered the loads, many of these included, in Franklin D. Richards's description, "cripples, and old grey-headed men." Historian's Office Minutes, 4 October 1856, Church Archives.

9. Stegner, *The Gathering of Zion;* and B. H. Roberts, *A Comprehensive History of the Church of Jesus Christ of Latter-day Saints: Century I,* 6 vols. (Salt Lake City: The Church of Jesus Christ of Latter-day Saints, 1930); 4:83–107. Both cite Chislett's reference to Levi Savage, subcaptain and one of the few dissenters in the decision to continue west that season. A lesser-known account of the "council meeting" in Florence at which the decision was made may be found in "Laletas [Laleta] Dixon's History of Her Ancestor, William James of Willey [Willie] Handcart Co., 1856," typescript, Church Archives. This history incorporates first-person reminiscences by Sarah, Emma, and Mary Ann James, who were eighteen, sixteen, and eleven years of age during the trek. Emma reports (p. 2) that as a result of this meeting one hundred Saints "decided to winter over," but the majority "voted to go on."

10. Although there has been more than adequate treatment of the handcart experience as a whole, the only extensive attention given to the relief effort was in 1914 when a series of articles by Solomon F. Kimball entitled "Belated Emigrants of 1856" was published in volume 12 (nos. 1–4) of *The Improvement Era.* These were partly based on Andrew Jenson's 1893 series in *The Contributor,* cited earlier. Kimball had the great advantage of personal acquaintance with many of the rescue party "heroes." The Hafens devote twenty-three pages to the rescue, relying principally on the Grant party journal kept by Robert T. Burton, as copied into the Journal History of the Church, Church Archives, under the date 30 November 1856. (The Journal History of the Church is hereafter referred to as JH, Church Archives.) See also John Chislett's account published in T. B. H. Stenhouse, *The Rocky Mountain Saints* (New York: D. Appleton, 1872), pp. 312–32; and the writings of Daniel W. Jones and Ephraim K. Hanks, cited below.

11. "Laleta Dixon's History," Church Archives, p. 4.

12. Levi Savage, Jr., Journal, entries for 8–18 October 1856, typescript, Church Archives.

13. Short histories of Mary Taylor, John Jaques, Josephine Zundle, and others, all members of the Martin company, may be read in Kate B. Carter, comp., *Treasures of Pioneer History,* 6 vols. (Salt Lake City: Daughters of Utah Pioneers, 1952–57), 5:229–96. See also the Journal of Thomas Durham, typescript, Church Archives.

14. JH, Church Archives, 24 and 27 September 1856. Roberts raises the question of what became of these supply wagons, which seem to have played no role in the rescue although they were counseled by Richards to go to the aid of the Willie company. Perhaps these teamsters (Patriarch John Smith with two others, and William Smith

with two men from Farmington) traveled east for several days but grew discouraged when they did not find the Willie company, which at that time was still more than 150 miles east on the Platte River. See Roberts, *Comprehensive History,* 4:90 fn.

15. These statistics are borrowed from Hafen and Hafen, *Handcarts,* p. 193; and Carter, *Treasures,* 6:44. Both drew from mission and immigration records and vary only slightly.

16. *JD,* 4:115.

17. Historian's Office Minutes, 4 October 1856, Church Archives. There is a pathetic note in Richards's conference report. Between 1846 and 1856 he had been assigned three missions to England, so that for seven of the ten years he was separated from his family. During his absences all three of his children died. Perhaps he had cause to expect others to make sacrifices for the "Kingdom."

18. Ibid.

19. Ibid.

20. Brigham Young to Amasa Lyman, 4 November 1856, Church Archives.

21. Brigham Young speech of 5 October 1856, reported in *Deseret News,* 15 October 1856.

22. Daniel W. Jones, *Forty Years among the Indians* (Salt Lake City: Juvenile Instructor's Office, 1890), p. 60. This Dan Jones, cook for the main rescue party, was born in 1830 in Missouri and baptized in 1851 after marching with the Mormon Battalion. He should not be confused with Dan Jones, born in 1811, who pioneered the Welsh Mission, returning from Europe in 1856.

23. Brigham Young's 5 October 1856 speech as reported in JH, Church Archives, for that date.

24. "Original Historical Narrative of Lucy Meserve Smith, Salt Lake City, Utah, August 14, 1888–89, written by Herself," p. 53, microfilm copy, Church Archives.

25. Jones, *Forty Years,* p. 60.

26. Journal of Harvey H. Cluff, Harold B. Lee Library, Brigham Young University, Provo, Utah, apparently written in 1868 from contemporary notes. The pertinent extract comprises Appendix G in Hafen and Hafen, *Handcarts,* pp. 232–38.

27. The more complete version of Hanks's handcart narrative was told to Andrew Jenson and published in *The Contributor* 14 (January 1893):201–5. The story is also told in S. A. Hanks and E. K. Hanks, *Scouting for the Mormons on the Great Frontier* (Salt Lake City: Deseret News Press, 1946). Hanks afterwards married, as a plural wife, Thisbe Read of the Martin company.

28. During the period 1850 to 1852 George D. Grant's company was mobilized six times, defending settlers in Utah County to the south, Skull Valley to the west, Shoshone lands in the north, and Green River to the east. Andrew Jenson, *Latter-day Saints' Biographical Encyclopedia,* 4 vols. (Salt Lake City: Andrew Jenson History Company, 1901–36), 1:239–40. For information on these militias and their mobilization, see Lawrence G. Coates, "Brigham Young and Mormon Indian Policies: The Formative Period, 1836–51," *Brigham Young University Studies* 18 (Spring 1978):428–52; and Howard A. Christy, "Open Hand and Mailed Fist: Mormon-Indian Relations in Utah, 1847–52," *Utah Historical Quarterly* 46 (Summer 1978):216–35; and Christy, "The Walker War: Defense and Conciliation as Strategy," *Utah Historical Quarterly* 47 (Fall 1979):395–420.

29. This information has been gathered from sketches in Jenson, *Biographical Encyclopedia.*

30. "The Autobiography of Asa Hawley," typescript, Church Archives, p. 3. Hawley, twenty-one at the time of the handcart rescue, wrote his story in 1912, possibly from an old diary.

31. Some of the Minute Men who joined early or later rescue parties were William K. Broomhead, George W. Grant, Asa Hawley, Benjamin Hampton, C. Allen Huntington, Heber P. and David P. Kimball, Ira Nebeker, Stephen W. Taylor, and Brigham Young, Jr. This list was compiled from Jones, *Forty Years*; "Autobiography of Asa Hawley," Church Archives; and Jenson, *Biographical Encyclopedia*. See also Solomon F. Kimball, "Our Pioneer Boys," *The Improvement Era* 11 (July 1908):668–74.

32. Kimball, "Our Pioneer Boys," p. 669.

33. This report was recorded in a ledger entitled "Freight Co. Account Book," now in the records of the North Thirteenth Ward, University West Stake, Church Archives. Unfortunately the report was undated. Thus, it is possible that the list pertains to a supply operation of another year's immigration or the Brigham Young Express and Carrying Company, established in 1856. Since Rumell was clerk between November 1852 and April 1857 (see Record of Members for the North Thirteenth Ward, microfilm, Church Archives), the note likely applies to an 1855 or 1856 operation. Whether or not it originated in the handcart rescue, the report reveals how one bishop responded to a similar church call for assistance.

34. Burton Journal, JH, Church Archives.

35. Tithing Office Account Book, 1856, Church Archives.

36. Jenson, "Belated Emigration," provides biographical information on several rescuers for whom no other information was found, either historical or genealogical.

37. For instance, Bishop Woolley of the Thirteenth Ward found George D. Grant an unsavory companion on a cattle drive to Sacramento. On another occasion a citizen wrote an anonymous note to Woolley complaining of Grant's and others' foul language in a street fight. See Leonard J. Arrington, *From Quaker to Latter-day Saint: Bishop Edwin D. Woolley* (Salt Lake City: Deseret Book Company, 1976), pp. 308, 331–32.

38. "Autobiography of Asa Hawley," Church Archives, p. 5.

39. For dates and details on this first leg of the relief excursion we have relied on the Burton Journal, JH, Church Archives; supplemented by Jones, *Forty Years*; Captain Grant to Brigham Young, 2 November 1856, written from Devil's Gate and published in the *Deseret News*, 19 November 1856; and Reddick N. Allred's diary printed in Carter, *Treasures*, 5:344–45.

40. Jones, *Forty Years*, p. 61.

41. Levi Savage, Jr., Journal, Church Archives, 19 October 1856; Burton Journal, JH, Church Archives, 13 October 1856.

42. Jones, *Forty Years*, p. 62.

43. Jenson, *Biographical Encyclopedia*, 1:240.

44. "Laleta Dixon's History," Church Archives, p. 4.

45. "Short [Autobiographical] History of Jens Nielson," Church Archives, p. [4].

46. "Laleta Dixon's History," Church Archives, p. 4.

47. Biography of Mary Hurren Wight, in JH, Church Archives, 8 November 1856.

48. "Laleta Dixon's History," Church Archives, p. 4.

49. John Oborn Autobiography in Kate B. Carter, ed., *Heart Throbs of the West*, 12 vols. (Salt Lake City: Daughters of Utah Pioneers, 1936–51), 6:364–66; and Stenhouse, *Rocky Mountain Saints*, pp. 312–32.

50. John Oborn Autobiography in Carter, *Heart Throbs*, 6:366; and John Chislett's narrative in Stenhouse, *Rocky Mountain Saints*, p. 322.

51. William Woodward Diary, in JH, Church Archives, 8 November 1856.

52. Harvey H. Cluff Journal, in Hafen and Hafen, *Handcarts*, p. 233.

53. Ibid.

54. Jones, *Forty Years*, p. 65.

55. Mary Hurren Wight Biography, in JH, Church Archives, 8 November 1856.

56. "Laleta Dixon's History," Church Archives, pp. 5–6.

57. Jones, *Forty Years,* p. 65.

58. Harvey H. Cluff Journal, in Hafen and Hafen, *Handcarts,* p. 234.

59. "Laleta Dixon's History," Church Archives, p. 6.

60. Ibid.

61. Ibid., p. 5.

62. Ibid.

63. Chislett's narrative in Stenhouse, *Rocky Mountain Saints,* pp. 330–31; Woodward Diary, and James G. Willie History, both in JH, Church Archives, 8 November 1856.

64. Woodward Diary, in JH, Church Archives, 8 November 1856.

65. Autobiography of Silas Richards in Kate B. Carter, comp., *Our Pioneer Heritage,* 20 vols. (Salt Lake City: Daughters of Utah Pioneers, 1958–77), 15:109–17.

66. Brigham Young to Amasa Lyman, 4 November 1856, Church Archives.

67. JH, Church Archives, 2 November 1856. Some of these, said Grant, were prepared with seven days' feed for their animals. "It will be necessary for more teams to go to their relief with grain and hay enough to sustain the animals already out, or they will die."

68. Juanita Brooks, ed., *On the Mormon Frontier: The Diary of Hosea Stout,* 2 vols. (Salt Lake City: Utah State Historical Society, 1964), 2:605, entry for 8 November 1856.

69. Woodward Diary; see also Levi Savage, Jr., Journal, 22–29 October 1856; and James G. Willie History; all three in JH, Church Archives, under the date 8 November 1856.

70. Jones, *Forty Years,* p. 64.

71. "Some historians give other names," Jones wrote of this second express, "but I was there myself and am not mistaken." He may have been referring to a statement made in an anonymous reminiscence of the Martin company experience published in Stenhouse, *Rocky Mountain Saints,* pp. 337–38.

72. George D. Grant Report to Brigham Young, 2 November 1856, Church Archives; also published in *Deseret News,* 19 November 1856.

73. Jones, *Forty Years,* p. 65.

74. Joseph A. Young speech given in Salt Lake Tabernacle, 16 November 1856, reported in *Deseret News,* 19 November 1856.

75. "In the morning we would find their starved and frozen bodies right by the side of us," Jane Griffiths Fullmer stated, "not knowing when they died until daylight revealed the ghastly sight to us." "An Early Pioneer History and Reminiscences written by Ella Campbell from material which she gathered from the Pioneers, February 5, 1914," photocopy of typescript in Church Archives, p. 3. A common observation by participants was that women seemed to have greater endurance and a lower fatality rate than men and boys. Joseph M. Simmons, member of Grant's party, in later years "often said that the women stood the suffering better by far than the men." Autobiography of Rachel Simmons in Carter, *Heart Throbs,* 11:168.

76. This description of the Willie company by George Cunningham applies equally to the Martin company. See Carter, *Treasures,* 5:254.

77. George D. Grant Report, 2 November 1856, Church Archives.

78. Story of Josiah Rogerson, *Salt Lake Tribune,* 14 January 1914.

79. Stenhouse, *Rocky Mountain Saints,* p. 324.

80. Stella Jaques Bell, ed., *Life History and Writings of John Jaques, including a Diary of the Martin Handcart Company* (Rexburg, Idaho: Ricks College Press, 1978), p. 149. This volume consists largely of Jaques's "Reminiscences," serialized in the *Salt Lake Daily Herald,* 1878–79, and selections from "Diary of Patience Loader Rosa [Rozsa] Archer," typescript, Harold B. Lee Library, Brigham Young University, Provo, Utah. Archer's holograph is entitled "Recollections of Past Days," referred to hereafter as Archer, "Recollections."

81. Jones, *Forty Years,* p. 65.

82. Ibid.

83. Ibid., p. 66.

84. Ibid.

85. Harvey H. Cluff Journal, in Hafen and Hafen, *Handcarts,* p. 235.

86. Autobiography of Heber Robert McBride, photocopy of typescript, Church Archives, p. 14. Young Heber reported unhappy encounters with Daniel Tyler, for one, who kicked over Heber's little tin can of hard-kindled coals when the McBrides did not gather one evening for camp prayers. Several authors mention instances of pettiness and cruelty, such as when a driver whipped a man who was holding onto a slowly moving wagon for support, and whipped a man who sneaked out one night to kill an abandoned cow, hiding the beef so that his family could eat better than the common rations. Overall, commented one writer, the instances of bravery—such as the fathers who gave up meals so their children would not be so hungry—far predominated.

87. Burton Journal, JH, Church Archives.

88. Harvey H. Cluff Journal, in Hafen and Hafen, *Handcarts,* p. 235.

89. Archer, "Recollections," p. 84. Patience was nineteen during the handcart experience. See also "Patience Loader Rozsa Archer," in Carter, *Our Pioneer Heritage,* 14:263.

90. Jones, *Forty Years,* p. 68.

91. Ibid.

92. George D. Grant to Brigham Young, 2 November 1856, Church Archives.

93. As told in Orson F. Whitney, *History of Utah,* 4 vols. (Salt Lake City: George Q. Cannon and Sons, 1892–1904), 1:555–65.

94. Autobiography of Heber Robert McBride, Church Archives, p. 14.

95. Archer, "Recollections," p. 85. Some punctuation supplied.

96. Jones, *Forty Years,* p. 68. Perhaps he referred to Andrew Jenson, who extolled the contribution of Ephraim K. Hanks and a few others.

97. Archer, "Recollections," p. 86.

98. Bell, *Writings of John Jaques,* p. 160.

99. Stephen Wells Taylor, twenty-three at the time, was a son-in-law of Brigham Young. David Patton Kimball, eighteen, who died at age forty-eight, was a son of Heber C. Kimball and Vilate Murray. George W. Grant, eighteen, was a son of Captain George D. Grant. C. Allen Huntington, eighteen, son of Dimick Huntington, also died young. Patience Loader wrote: "This poor Brother Kimble [David Kimball] staid so long in the water that he had to be taken out and packed to camp and he was a long time before he recovered as he was chilled through and in after life he was always afflicted with rheumatism." Archer, "Recollections," pp. 86–87.

100. Burton Journal, JH, Church Archives, 9 November 1856. Jones leaves a harrowing account of their winter at Devil's Gate. Their ordeal was terrible. Their cattle died, they could not find wild game to kill, and they subsisted on hide from the dead cattle and coffee from the stores for six weeks. An occasional saddle moccasin and wagon tongue cover was boiled to make a broth. "We asked the Lord to bless our stomachs and adapt them to this food," wrote Jones. "We hadn't the faith to ask him to bless the raw-hide for it was 'hard stock.'" Eventually, they were "saved" by buffalo meat traded them by friendly Indians and by flour left with them in early spring by advance parties of the Brigham Young Express and Carrying Company, established by the church to carry mail, freight, and passengers to and from the Missouri River. See Jones, *Forty Years,* pp. 60–119.

101. Jenson, "Belated Emigration," p. 203.

102. Ibid.

103. Burton Journal, JH, Church Archives, 12 November 1856.

104. Brigham Young speech of 4 November 1856, recorded in *Deseret News,* 12 November 1856.

105. Brigham Young Office Journals, 1852–57, Church Archives, 4–17 October 1856.

106. Ibid., 18–31 October 1856.

107. Ibid., 31 October 1856.

108. Autobiography of Silas Richards, in Carter, *Our Pioneer Heritage,* 15:109–17.

109. Diary of Goudy Hogan, holograph in Joel E. Ricks Collection, Utah State University Library, Logan, Utah.

110. "Autobiography of Asa Hawley," Church Archives, p. 6.

111. Brooks, *Diary of Hosea Stout,* 2:605, 11 November 1856.

112. Ibid., pp. 605–6, 12 November 1856.

113. Ibid., p. 606, 17 November 1856.

114. Ibid., 18 November 1856.

115. Ibid., p. 607, 19 November 1856.

116. "Autobiography of Asa Hawley," Church Archives, p. 6.

117. Photocopy of handwritten letter in Joseph Simmons Willes, "A Brief Biography of Joseph Marcellus Simmons," typescript, with photographs, 31 pp., Church Archives. At the time, Joseph Simmons, son-in-law of Brigham Young's business manager, was twenty-two years of age. His first child was born in Salt Lake City while he was on the rescue mission. See Carter, *Heart Throbs,* 11:168.

118. Brooks, *Diary of Hosea Stout,* 2:607, 20–25 November 1856.

119. Brigham Young Office Journal, Church Archives, 25 November 1856.

120. Burton Journal, JH, Church Archives, 30 November 1856; Bell, *Writings of John Jaques,* pp. 166–71; "Autobiography of Asa Hawley," Church Archives, p. 6; and Brooks, *Diary of Hosea Stout,* 2:608, 30 November 1856.

121. Brigham Young Office Journal, Church Archives, 15 December 1856.

122. Journal of William Wight, clerk of D. D. McArthur Handcart Company, unpaged, Church Archives, 26 September 1856.

123. Bell, *Writings of John Jaques,* p. 172.

124. Autobiography of Silas Richards, in Carter, *Our Pioneer Heritage,* 15:114.

125. "Narrative of Lucy Meserve Smith," Church Archives, p. 53.

126. "Emigration East," entries for 6 October and 15 December 1856, Tithing Office Account, Church Archives.

127. "Narrative of Lucy Meserve Smith," Church Archives, p. 54.

128. JH, Church Archives, 8 November 1856, p. 1.

129. Campbell, "An Early Pioneer History," Church Archives.

130. Carter, *Treasures,* 6:53.

131. Carter, *Heart Throbs,* 6:378.

132. Bell, *Writings of John Jaques,* p. 173.

133. Carter, *Treasures,* 5:255.

134. Carter, *Heart Throbs,* 6:366.

135. Archer, "Recollections," p. 92.

136. Autobiography of Silas Richards, in Carter, *Our Pioneer Heritage,* 15:115.

137. Ibid., 12:295.

138. Autobiography of Heber Robert McBride, Church Archives, p. 16.

139. Campbell, "An Early Pioneer History," Church Archives, pp. 3–4.

140. Speech of 2 November 1856, *JD,* 4:62.

141. Brigham Young speech in Tabernacle, 2 November 1856, published in *Deseret News,* 12 November 1856.

142. Stenhouse, *Rocky Mountain Saints,* pp. 332, 339.

143. Ibid., p. 332; Woodward Diary, in JH, Church Archives, 8 November 1856.

Index